Affinity Biosensors

METHODS IN BIOTECHNOLOGY™

John M. Walker, SERIES EDITOR

METHODS IN BIOTECHNOLOGY™

Affinity Biosensors

Techniques and Protocols

Edited by

Kim R. Rogers

US-EPA, Las Vegas, NV

and

Ashok Mulchandani

University of California, Riverside, CA

Humana Press ✳ Totowa, New Jersey

Cover illustration: Fig. 2 in Chapter 10, "Immunobiosensors Based on Ion-Selective Electrodes," by Hanna Radecka and Yoshio Umezawa

Cover design by Patricia F. Cleary.

For additional copies, pricing for bulk purchases, and/or information about other Humana titles, contact Humana at the above address or at any of the following numbers: Tel.: 973-256-1699; Fax: 973-256-8341; E-mail: humana@humanapr.com, or visit our Website: http://humanapress.com

Printed in the United States of America. 10 9 8 7 6 5 4 3 2 1

Library of Congress Cataloging in Publication Data

Main entry under title:

Methods in biotechnology™.

Affinity biosensors: techniques and protocols/edited by Kim R. Rogers and Ashok Mulchandani.
    p.   cm.—(Methods in biotechnology; 7)
    Includes bibliographical references and index.
    ISBN 0-89603-539-5 (alk. paper)
    1. Biosensors. 2 Immunotechnology. I. Rogers, Kim R., 1956– . II. Mulchandani, Ashok, 1956– . III. Series.
R857.B54A395   1998
612'.015'028—dc21   98-4566
                                                                            CIP

Preface

The frequency of reports concerning the interface of biological recognition elements to signal transduction technologies has risen dramatically over the last decade. Because any one of a wide variety of biological recognition elements (e.g., antibodies, receptors, DNA, microorganisms, or enzymes) can theoretically be interfaced with any one of a wide variety of signal transducers (e.g., optical, electrochemical, thermal, or acoustic), the potential range of devices and techniques can be bewildering. The purpose of this volume and the previous volume in this series is to provide a basic reference and starting point for investigators in academics, industry, and government to begin or expand their biosensors research. This volume, *Methods in Biotechnology, vol. 7: Affinity Biosensors: Techniques and Protocols*, describes a variety of classical and emerging transduction technologies that have been interfaced to bioaffinity elements (e.g., antibodies and receptors).

Some of the reasons for the expansion in the use of affinity-based biosensors include both advances in signal transduction technologies (e.g., fiber optics, microelectronics, and microfabrication) and the availability of bioaffinity elements. More specifically, with respect to biological recognition elements, commercially and noncommercially produced antibodies directed toward a variety of analytes have become widely available. In addition, techniques for the purification and stabilization of receptors have also significantly improved. As a result of these recent advances in the field, biosensors research and development projects are being pursued by investigators from a wide range of disciplines.

There are a variety of approaches that researchers employ to select a combination of bioaffinity elements and signal transducers. One commonly used approach is to identify the compound or compounds of interest, identify the bioaffinity molecule that yields an appropriate selectivity and dynamic concentration range for the assay, and choose an assay format and signal transduction technology that will meet the analytical requirements for the proposed application. This volume, although not an exhaustive treatise, provides a detailed "step-by-step" description for a variety of affinity-based biosensor technologies that will allow the novice or experienced investigator to expand into new areas of research most appropriate for their analytical needs.

v

This volume is divided into two sections, covering affinity biosensors and biosensor-related techniques. Chapter 1 provides an overview of the principles relevant to the design and operational features of affinity-based biosensors and related techniques. The subsequent chapters in the first section provide detailed protocols for affinity biosensors based on optical, electrochemical, thermal, acoustic, and plasmon resonance techniques. Primarily as a result of the widespread use of antibody-based assays, immunochemical biosensors represent the largest group for affinity biosensors. Included in the second section are techniques for which the biological recognition element is intimately involved but not necessarily directly interfaced to the signal transducer. Because biosensors are centrally located in a continuum of analytical methods, a variety of biosensor-related techniques will continue to impact the more classical concepts of biosensors.

In addition to the detailed protocols, special notes and safety items have been included to provide details that may not normally appear in journal article descriptions. These notes can be particularly useful for those not familiar with construction and operation of a specific device or technique.

We are fortunate to have assembled contributions from world-class authorities in this field and we sincerely thank them. In their enthusiasm for the field of biosensors research they have produced volumes that we believe will be of unusual help to the increasing number of researchers in this field. We are indebted to Prof. John Walker, Series Editor, for his careful attention in reviewing the manuscripts included in this volume. Last but not least, we warmly acknowledge the gracious support of our families.

Kim R. Rogers
Ashok Mulchandani

Contents

Contents for the companion volume:
Enzyme and Microbial Biosensors

ix

Contributors

F. PHILIP ANDERSON • *Section of Clinical Chemistry, Department of Pathology, Medical College of Virginia, Virginia Commonwealth University, Richmond, VA*

J. REX ASTLES • *Division of Laboratory Systems, Public Health Practice Program Office, Centers for Disease Control and Prevention, Atlanta, GA*

RALPH BALLERSTADT • *Center for Biotechnology and Bioengineering, University of Pittsburgh, PA*

PIET BERGVELD • *MESA Research Institute, University of Twente, Enschede, The Netherlands*

GEERT A. J. BESSELINK • *MESA Research Institute, University of Twente, Enschede, The Netherlands*

FRANK BIER • *Max-Delrück Center for Molecular Medicine, University of Potsdam, Berlin, Germany*

URSULA BILITEWSKI • *GBF, Braunschweig, Germany*

ALBRECHT BRANDENBERG • *Fraunhofer Institute for Physical Medicine, Freiburg, Germany*

CHENG J. CAO • *Department of Pharmacology and Experimental Therapeutics, University of Maryland, Baltimore, MD*

ROBERT CARTER • *Universal Sensor Inc., Metaire, LA*

VANIA I. CORTES • *Laranjeiras, Brazil*

BENGT DANIELSON • *Department of Pure and Applied Biochemistry, University of Lund, Sweden*

RICHARD A. DURST • *Analytical Chemistry Laboratories, Department of Food Science and Technology, Cornell University, Geneva, NY*

MOHYEE E. ELDEFRAWI • *Department of Pharmacology and Experimental Therapeutics, University of Maryland School of Medicine, Baltimore, MD*

AMIRA T. ELDEFRAWI • *Department of Pharmacology and Experimental Therapeutics, University of Maryland School of Medicine, Baltimore, MD*

GEORGE G. GUIBAULT • *Department of Chemistry, University College, Cork, Ireland*

C. MICHAEL HANBURY • *Roche Diagnostics, Somerville, NJ*

AYUMI HIRANO • *Department of Chemistry, School of Science, University of Tokyo, Japan*

†John J. Horvath • *Biotechnology Division, National Institute of Standards and Technology, Gaithersburg, MD*

Knut Johansen • *Department of Physics, IFM, Linkoping University, Linkoping, Sweden*

Masoud Khayyami • *Department of Pure and Applied Biochemistry, University of Lund, Sweden*

Bo Liedberg • *Department of Physics, IFM, Linkoping University, Linkoping, Sweden*

Frances S. Ligler • *Center for Bio/Molecular Science and Engineering, Naval Research Laboratory, Washington, DC*

Darrel E. Menking • *SCBRD-RT, ERDEC, United States Army, Aberdeen Proving Ground, MD*

Marco Mascini • *Dipartimento di Sanita Pubblica Epidemiologic e Chimica Analitica Ambientale, University Delgi Studi di Firenze, Italy*

W. Greg Miller • *Section of Clinical Chemistry, Department of Pathology, Medical College of Virginia, Virginia Commonwealth University, Richmond, VA*

Maria Minunni • *Dipartimento di Sanita Pubblica Epidemiologic e Chimica Analitica Ambientale, University Delgi Studi di Firenze, Italy*

Robert J. Mioduszewski • *SCBRD-RT, ERDEC, United States Army, Aberdeen Proving Ground, MD*

Hanna Radecka • *Department of Chemistry, School of Science, University of Tokyo, Japan*

Kumaran Ramanathan • *Department of Pure and Applied Biochemistry, University of Lund, Sweden*

Matthew A. Roberts • *Department De Chimie-ICP III, Laboratoire de Electrochimie, Lausanne, Switzerland*

Kim R. Rogers • *US Environmental Protection Agency, Las Vegas, NV*

Jerome S. Schultz • *Center for Biotechnology and Bioengineering, University of Pittsburgh, PA*

Masao Sugawara • *Department of Chemistry, School of Science, University of Tokyo, Japan*

Yoshio Umezawa • *Department of Chemistry, School of Science, University of Tokyo, Japan*

James J. Valdes • *SCBRD-RT, ERDEC, United States Army, Aberdeen Proving Ground, MD*

Randy M. Wadkins • *Center for Bio/Molecular Science and Engineering, Naval Research Laboratory, Washington, DC*

†Deceased.

I

AFFINITY BIOSENSORS

1

Principles of Affinity-Based Biosensors

Kim R. Rogers

1. Introduction

The use of antibodies as recognition elements in bioanalytical assays can be traced back to the late 1950s. Yalow and Berson *(1)* pioneered radioimmunoassay for measurement of insulin and in the following year Ekins *(2)* used this technique to measure thyroxine. Reports involving the use of antibodies in devices that might now be referred to as biosensors began to emerge in the early 1970s with the work of Kronick and Little *(3)*, Giaever *(4)*, and Tromberg et al. *(5)*. In the past decade, a wide variety of affinity-based biosensor configurations have been reported. Current consensus opinion would suggest that affinity-based biosensors are analytical devices that use an antibody, sequence of DNA, or receptor protein interfaced to a signal transducer to measure a binding event.

Theoretically and as evidenced in the literature, affinity proteins can be interfaced to any number of signal transducers provided a mechanism of transduction can be designed. Different types of affinity-based sensor configurations have been pioneered by various research groups. This chapter is not intended to provide a comprehensive review of affinity-based biosensors, but rather to discuss the fundamental considerations of design and operation and place in perspective the protocols presented in this volume. The affinity-based biosensor section collects together some of the classical configurations as well as some more recent developments and provides a starting point for further research into the application of these techniques to specific analytical problems. Several current reviews may provide additional information on this subject *(6–9)*.

In the design and construction of a biosensor, a number of characteristics are important in the determination of whether a biosensor may be suitable for a particular application. These characteristics that together define the design and

From: *Methods in Biotechnology, Vol. 7: Affinity Biosensors: Techniques and Protocols*
Edited by: K. R. Rogers and A. Mulchandani © Humana Press Inc., Totowa, NJ

**Table 1
Fundamental Design and Operational Considerations
for Affinity-Based Biosensors**

Structural and design considerations	Operational considerations
Bioaffinity element properties	Sensitivity, selectivity, kinetic parameters, stability
Assay format	Homogeneous vs heterogeneous reversible, regenerable, disposable continuous, remote, *in situ* operation assay time
Sensor material	Immobilization method
Transducer type	Mechanism of signal transduction

operational characteristics of affinity-based biosensors include properties of the bioaffinity element, sensor surface material, method of immobilization, assay format, type of signal transducer, and mechanism of signal transduction (**Table 1**).

The binding of an analyte to a bioaffinity-based biosensor is stoichiometric in nature. Consequently, factors involved with the binding event are of particular importance to these biosensors. For example, the kinetics parameters of the antibody–antigen or receptor–ligand binding are critical to the design and operation. In addition, because there are a finite number of binding sites on the sensor surface, techniques used for immobilization of the recognition protein also become an important factor in the design, construction, and operation of these biosensors. Immobilization issues are particularly relevant for biological receptors, which are typically isolated from a cellular membrane environment. Signal transduction also plays an important role. Transducers that have been shown to be particularly useful in combination with bioaffinity recognition elements include optical, electrochemical, and piezoelectric devices.

2. Affinity Recognition Elements

A wide variety of bioaffinity elements have been used in biosensors. These include antibodies, receptors, biomimetic materials, and nucleic acids (**Table 2**). Affinity-based biosensors have been demonstrated to measure many of the same compounds for which bioanalytical assays have been previously developed. For example, analytes for antibody-based biosensors include low-mol-wt compounds *(7)*, proteins, and microorganisms *(8)*. Compounds that have been measured using receptor-based biosensors are similar to those using radioisotope-labeled ligand assays. These compounds are typically ligands of physiological, pharmacological, or toxicological significance with respect to

Table 2
Bioaffinity Elements for Affinity-Based Biosensors

Bioaffinity element	Types of analyte	Examples
Antibodies	Low mol-wt compounds	Drugs, hormones, environmental pollutants (pesticides, explosives, and so forth)
	Proteins	Antipathogen antibodies
		Toxins, insulin, serum proteins
	Microorganisms	*Candida albicans, Escherichia coli, Salmonella typhimurium, Salmonella dysenteria, Yersinia pestis*
Biological receptors		
Interleukin-6 receptor	Physiological ligands	
Acetylcholine receptor	Pharmacological ligands	Nicotine, carbamyl choline
	Toxicological ligands	Bungarotoxin
Nucleic acids	Identification of specific sequences	*Legionella pneumophila*
	Detection of intercalators	Ethidium, PAHs

5

the receptor. Biosensors for nucleic acids show considerable potential but have thus far been limited to the detection of hybridization of complimentary oligonuleotides *(10,11)* and intercalation of optically or electrochemically detectable compounds *(12)*.

Owing primarily to their high affinity, versatility, and commercial availability, antibodies are the most widely reported biological recognition elements used in affinity-type biosensors. Although antibodies of the IgG class are, to a certain extent, similar in structure, their affinities for antigens may vary widely. These differences in antigen specificity and binding affinity originate from the variations in amino acid sequence at the antigen binding site *(13)*. The antibody binding sites are located at the ends of two arms (Fab units) of this "Y-shaped protein." The base of the "Y" referred to as the Fc unit is less variant and contains species-specific structure that is commonly used as an antigen for production of species-specific (anti-IgG) antibodies. The differences in antibody–antigen binding characteristics influence the wide range of detection limits observed for antibody-based biosensors.

Antibodies are generated in response to the challenge of an immunogen in the host animal. Because small-mol-wt compounds (the analytes of interest, in many cases) are not themselves immunogenic, they are typically bonded to large-mol-wt proteins, such as bovine serum albumin or keyhole limpet hemocyanin. Antibodies derived from the serum of an immunized animal form an array of molecular populations (each arising from a separate cell line) that recognize various regions (haptens) on the immunogen. These antibodies are termed polyclonal. Antibodies that are derived from a single cell line (termed monoclonal) typically recognize a more specific region of the immunogen than do polyclonal populations.

The use of both polyclonal serum and monoclonal antibodies (MAbs) have certain advantages and limitations for use in biosensors. One of the most obvious issues relates to the density of binding sites that can be immobilized on the surface of the signal transducer. In this respect, MAbs offer some advantage because of the absence of serum proteins and other nonanalyte-specific antibodies. Nevertheless, a variety of innovative antibody-based biosensor formats have been reported that facilitate the concentration (on the sensor surface) of the IgG fraction of polyclonal antiserum *(14)*.

Reports of receptor-based biosensors as compared to antibody-based biosensors have been far less numerous in the literature. This is most likely the result of several complicating issues. The isolation and reconstitution of biological receptors is typically difficult and often requires specialized techniques. These proteins are usually quite labile, and because many receptors are isolated from biological membranes, they often require reconstitution into a lipid environment to express their physiological function.

One of the most studied and perhaps best understood receptor proteins is the nicotinic acetylcholine receptor (nAChR) *(15)*. Reasons why this receptor is attractive for use in biosensors (*see* Chapter 9) include its interaction with a wide variety of drugs and toxins, as well as the fact that it can be easily isolated in milligram quantities from the electric organ of *Torpedo nobilana (16)*. Although some receptors that have been incorporated into biosensors (*see* Chapter 12), such as protamine (a polypeptide that specifically binds to heparin), are commercially available, most receptor proteins that have been used for biosensor-related techniques are not. Biological receptors, such as the glutamate receptor ion channel and Na^+/glucose cotransporter, must be isolated by the research investigator from such sources as rat brain or guinea pig intestine, respectively (*see* Chapter 14; *17,18*).

3. Immobilization of Bioaffinity Elements

Owing to the stoichiometric relationship between affinity elements and ligands and the finite surface area of the signal transducer, immobilization and orientation (i.e., accessibility of the ligand binding site to the analyte) are important considerations in the design and construction of affinity-based biosensors. Although relatively few approaches have been explored for receptor proteins, a wide variety of methods have been reported for immunochemicals *(19)*. Immobilization approaches used for antibodies include covalent binding, entrapment, crosslinking, adsorption, and the use of biological binding proteins, such as protein A or protein G, or use of the avidin/biotin system.

Covalent immobilization typically involves modification of the sensor surface with activated compounds that react with various groups on the proteins, such as amines, hydroxyls (oxidized to formyl groups with sodium periodate), and sulfhydryls. This approach has been used to immobilize complete antibodies as well as isolated Fab units *(14,20)*. Immobilization of these antigen-binding portions of the antibody has been shown to increase the density of the binding sites as compared to the use of the complete IgG.

Immunochemicals have also been entrapped in such materials as polyacrylamide, polyvinyl alcohol, polyvinyl chloride, epoxy, or sol-gels *(9)*. Although protein entrapment methods allow for optimization of variables, such as protein loading, material density, and pore size, concerns for these methods include protein leakage, orientation of the antibody, and maintenance of biological activity. Another immobilization method that is routinely used for antibody and receptors is physical adsorption. This method is relatively simple and has been shown to be satisfactory for a limited (usually single use) number of assays using the same sensor.

One of the major limitations with the previously mentioned methods for antibody immobilization is the random orientation of the IgG molecule such that the antigen-binding domains on a significant portion of the immobilized antibody are inaccessible to the antigen. Methods that have been employed to optimize antibody orientation include the use of avidin/biotin, protein A (or protein G), and species-specific anti-IgG "capture" antibodies. In the case of the avidin/biotin system, avidin (a protein isolated from egg whites) is typically immobilized to the sensor using one of the previously mentioned methods. Biotin (a low-mol-wt cofactor that binds to avidin with high affinity) is typically bound to the Fc region of the antibody *(21)*. In addition to this commonly used format, a variety of other antibody immobilization schemes have been used with this versatile system.

Another method used to orient antibodies on the sensor surface is through the use of protein A or protein G. These polypeptides bind to the Fc region of antibodies leaving the antigenic site free of stearic hindrance. Again, these antibody binding proteins can be immobilized to the sensor using previously mentioned methods. Although one may encounter problems in the orientation of the binding site of protein A, it has been shown that the use of this method can significantly increase the antigen-binding capability per microgram of IgG immobilized to the sensor *(14)*.

Species-specific anti-IgG antibodies directed toward the Fc portion of IgG have also been used to orient the anti-analyte antibody onto the sensor surface. For this method, the anti-IgG is immobilized by one of the previously outlined methods. Then the anti-analyte antibodies are allowed to bind the anti-IgG coated surface. As a result of this procedure, the anti-analyte antibodies are oriented on the surface of the sensor with their antigen-binding sites facing the solution. This procedure can also be used to concentrate IgG from antiserum onto a defined surface *(22)*.

In comparison to antibodies, receptor proteins are considerably more diverse in both structure and function. Depending on the receptor assay requirements, these proteins can be immobilized by covalent binding (to a chemical group at the sensor surface), adsorption, or reconstitution into a bilayer lipid membrane (BLM) *(17,18,23)*.

Transmembrane and membrane-associated receptors can be solubilized, i.e., removed from their cellular membranes, purified, and stabilized in solution; in many cases, they retain their ability to bind ligands *(16)*. However, in cases in which the receptor transport function is the desired result, reconstitution into a surrogate of its membrane environment, e.g., a BLM, is required. For ion channel receptors, this is typically accomplished using a BLM formed across a small polished hole in a Teflon membrane that separates two solution chambers. Although reconstitution and electrochemical detection of receptor protein function is a classical technique practiced in neuropharmacology labs for several decades, steady advances in electrochemical techniques and supported (on a solid surface) BLMs have spurred interest in receptor-based biosensors research *(24)*.

4. Format Considerations

4.1. Antibody-Based Biosensors

Antibody-based biosensors are designed in a variety of ways, but generally fall into one of three basic configurations involving an immobilized antibody (Ab), an immobilized antigen (Ag), or antibodies and antigens that are not immobilized but rather confined in dialysis tubing attached to the end of an optical fiber. For the immobilized antibody formats, the antigen may be unlabeled (in cases where the binding can be directly detected), labeled with an optically or electrochemically active tracer, or labeled with an enzyme. Although a great deal of variation is possible within these basic formats, the type of signal transduction mechanism typically dictates to a significant extent the type and characteristics of the assay format.

Because most target analytes do not show any chemical or physical characteristics that easily differentiate them from other compounds in the mixture, most affinity-based biosensors depend on a competition assay format using analyte tracers;

$$Ab + Ag + Ag^* \rightarrow AbAg + AbAg^* \tag{1}$$

where Ag^* is the analyte tracer, AbAg is the antibody-antigen complex, and $AbAg^*$ is the antibody antigen-tracer complex. The tracer may be optically or electrochemically detectable; or an enzyme may be used that catalytically converts a substrate to a product that is detected by the transducer. In any case, the signal transduction mechanism must be able to differentiate between the relative amount of antibody-binding sites that are occupied by the analyte and analyte tracer.

4.1.1. Signal Transducer Considerations

A number of different types of signal transducers have been interfaced with bioaffinity recognition elcments (**Table 3**). These transducer mechanisms include optical, electrochemical, thermal, and acoustic. Because analytes of interest do not typically show physical properties that can be detected by the signal transducer, most affinity-based biosensors rely on a competitive assay format. These competitive formats may either be configured as direct or indirect assays. For the direct assay format, either the antibody is immobilized to the sensor surface followed by the direct detection of the binding of labeled analyte tracer, or the antigen is immobilized to the surface followed by the detection of antibody binding to the surface. By contrast, the indirect assay format requires the use of an enzyme-labeled analyte tracer. After the initial binding event, the unbound enzyme tracer must be washed from the sensor surface before adding of the enzyme substrate. Appearance of the catalytic product is then detected by the signal transducer.

Table 3
Signal Transducers for Affinity-Based Biosensors

Transducer type	Assay format[a]
Optical	
Fluorescence energy transfer	Direct
Bioluminescence	Indirect
TIRF[b]	Direct
SPR[c]	Direct
Grating coupler	Direct
Electrochemical	
Potentiometric	Indirect, direct
Amperometric	Indirect
Conductimetric	Indirect
Thermal	Indirect
Acoustic	
QCM[d]	Direct

[a]Direct and indirect formats are defined in the text.
[b]Total internal reflectance fluorescence.
[c]Surface plasmon resonance.
[d]Quartz crystal microbalance.

A number of reported antibody-based biosensors depend on the use of a labeled analyte tracer (**Fig. 1A**) that competes with unlabeled analyte for antibody (binding sites) immobilized to the sensor surface. The primary requirement for these assay formats is that the antibody tracer complex be differentiated from free tracer. Although this can be accomplished by washing the unbound tracer from the sensor surface, several techniques do not require this separation step. Optical methods that have been particularly useful in this regard use total internal reflectance fluorescence (TIRF) techniques.

TIRF is a sensitive technique that has been reported for a variety of biosensor applications *(25)*. This method has been used to measure the binding of antigens to immobilized antibodies (*see* Chapter 5), ligands to receptors (*see* Chapter 9 and **ref.** *23)*, and intercalators to DNA *(12)*. Because the volume of the evanescent zone surrounding the waveguide is relatively small as compared to that of the bulk solution, only the tracer that is bound to the surface-immobilized receptor is detected. Consequently, the requirement that the antibody–tracer complex be differentiated from the free tracer can be accomplished without a separation step.

Several signal transducers, such as surface plasmon resonance (SPR), grating couplers, and acoustic transducers, can directly measure antibody–antigen binding at the surface of the transducer. Although these techniques do not

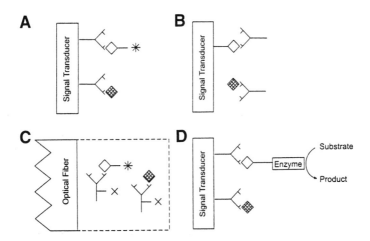

Fig. 1. Assay formats for bioaffinity-based biosensors. Schematic representation of the following formats: **(A)** direct-competitive based on immobilized antibody; **(B)** direct-competitive based on immobilized antigen; **(C)** direct-competitive based on fluorescence energy transfer; and **(D)** indirect-competitive based on immobilized antibody with an enzyme-tracer.

require the use of a fluorescent or enzyme-labeled tracer, they do typically require the use of a format in which large-mol-wt molecules, i.e., antibodies, bind to an immobilized antigen (**Fig. 1B**). Consequently, for small-mol-wt analytes, these assays operate in a competitive mode in which the signal is inversely related to the analyte concentration. For these formats, the immobilized antigen may be thought of as the antigen tracer Ag* (*see* **Eq. 1**).

SPR-based biosensors essentially measure the difference in refractive index between a protein and the water it displaces on binding to the sensor surface. Because this technique is highly versatile in measuring the real-time binding of proteins to surface immobilized ligands, it has been widely explored for numerous applications (*see* Chapter 3 and **ref. 26**). One of the features that makes this transduction technique attractive for use in affinity-based biosensors is the versatile coupling chemistry and kinetics software that are commercially available. Because SPR is sensitive to refractive index changes extending 200–300 nm away from the surface, carboxymethyl dextran chains with multiple coupling sites have been used to increase the density of immobilized antigen or ligand on the sensor surface. In addition, real-time analysis allows the determination of association and dissociation constants under appropriate conditions.

The use of grating couplers is another technique that has provided the basis for a variety of biosensors used to study bioaffinity. This method also measures changes in the refractive index at the sensor surface and has been used to measure binding of antibodies to antigens (*see* Chapter 8 and **ref. *27*)**, receptors to ligands, or DNA hybridization *(28)*. Similar to SPR, ligands or antigens can be immobilized to the grating coupler using a variety of techniques. These methods typically involve the use of aminopropyltriethoxy silane (APTES) to functionalize the surface followed by direct coupling of the protein or indirect immobilization of the antibody using protein A or avidin/biotin.

Acoustic wave transducers have also been widely used to detect antibody binding to an immobilized antigen (*see* Chapter 7 and **ref. *29*)** or high-mol-wt antigens binding to immobilized antibodies *(30)*. The simplest, least expensive, and most widely used acoustic transducer is the quartz crystal microbalance (QCM), also termed thickness-shear-mode sensor. This technique measures small changes in surface properties, such as bound surface mass and surface viscocity, resulting from the binding of high-mol-wt molecules to the sensor surface.

A variety of methods have been reported for immobilization of antigen or antibody to the surface of the sensor. Similar to the grating coupler, the most common technique for the QCM involves surface modification with APTES followed by coupling of the antibody or antigen using a linker, such as glutaraldehyde.

In another type of biosensor format, optical waveguides are used to interrogate immunoassays located a distance from the spectrometer (*see* Chapter 7). For many of these formats, both the antibody or receptor and the antigen or ligand are labeled with fluorescent tracers. In this configuration, the optical waveguide is used as a means to transfer light between the spectrometer and the binding reaction. These binding assays typically rely on fluorescence energy transfer (**Fig. 1C**). Fluorescence energy transfer requires that the emission spectrum of the donor (fluorescent tracer) significantly overlaps the excitation spectrum of the acceptor (fluorescent tracer). When the shorter wavelength tracer is excited, the binding reaction can be monitored by a decrease in the observed fluorescence for the donor or an increase in the fluorescence for the longer wavelength acceptor.

In contrast to the measurement of the direct binding of an antibody to an antigen or an antigen-tracer, a number of biosensor-related methods use indirect formats that rely on enzymatic tracers. In these formats, the enzyme-labeled analyte–tracer and the analyte compete for available binding sites on the immobilized antibody. Following separation of the unbound tracer from the sensor using a washing step, an enzyme substrate is added. The conversion of substrate to product is then used to measure the relative amount of antibody-binding sites occupied by the enzyme-labeled analyte tracer (**Fig. 1D**).

As a result of the versatility of enzyme tracers as well as the physical and chemical properties of the enzyme substrates and products, a variety of signal transducers have been used in this type of assay. Indirect antibody-based sensors include electrochemical (e.g., potentiometric, conductimetric, and amperometric) *(9)*, bioluminescent *(31)*, and calorimetric (*see* **ref.** *32* and Chapter 2) techniques. Because of the number and complexity of steps and manipulations required, however, this type of assay format sacrifices many of the potential advantages of the biosensors that use direct assay formats.

Electrochemical transducers are the best characterized and perhaps the easiest to commercialize. Although affinity-based assays do not easily lend themselves to direct electrochemical detection, a number of indirect configurations have been reported *(9)*. For example, enzyme-catalyzed reactions that result in a change in the concentration of carbon dioxide or an ionic species can be potentiometrically determined; the catalytic formation of electrochemically active products (e.g., H_2O_2, phenols) can be measured amperometrically *(33)* or conductimetrically *(34)*. Although these assay formats catalytically amplify the stoichiometric binding, they require separation and incubation steps that add additional time and complicate the assay. In addition to the indirect antibody based electrochemical biosensors, several techniques have been recently reported that directly measure the binding of an analyte to an immobilized antibody *(35)*. These biosensor methods appear promising for the detection of both electrochemically active and nonactive compounds.

4.2. Receptor-Based Biosensors

Membrane-bound receptor proteins typically bind to a small-mol-wt ligand that triggers or facilitates their physiological response. Receptors, such as the nAChR, that have been isolated and reconstituted into artificial membrane systems are assayed by either ligand-binding assays or by ligand-induced functional assays *(15)*. This is also the case for receptor-based biosensors and biosensor-related techniques. For example, the nAChR has been reported as a biological recognition element for biosensors that measure either ligand-induced biological function (i.e., ion movement through a membrane-bound channel) *(36)* or the ability to bind several classes of bioactive drugs and toxicants *(37)*.

In addition to binding functions, purified and reconstituted membrane receptors, such as the glutamate receptor ion channel and Na^+/glucose cotransporter, show ligand-induced transport functions. For many of these proteins, the binding of a ligand results in a conformational change in the protein that allows the movement of a compound or ion down its concentration gradient or the cotransport of a small-mol-wt molecule with an ion *(17)*. Because thousands of ions may move across the BLM as a result of the binding of one ligand, this type of assay shows the potential to be very sensitive. These biosensor and biosensor-related methods provide a basis for mechanistic studies into the function of these proteins and show the potential to provide biosensors based on neurophysiological and pharmacological function.

5. Operational Considerations

In addition to structural and design considerations, several operational issues must be considered in the design of affinity-based biosensors (**Table 1**). Bioaffinity elements used in biosensors are typically well characterized (in terms of solution-based assays) and are not expected to behave substantially differently on incorporation into biosensors. Considerable immobilization method-dependent variations, however, have been reported in both the amount of antibody immobilized to the sensor and the retention of binding capacity *(14)*. In addition, kinetics parameters for binding of an analyte to an immobilized affinity element are expected to be modified from those measured in solution *(25)*.

Heterogeneous operation refers to the requirement for separation of the bound and unbound analyte-tracer prior to assessment of the relative percentage of bound tracer. By contrast, homogeneous assays do not require this separation step. Advantages for homogeneous assays primarily involve their use in portable, remote, or *in situ* formats. However, these advantages become less compelling for biosensors that employ flow systems, since separation of the unbound tracer is easily facilitated as part of the assay.

Another operational consideration relevant to biosensor design involves the reversibility of the assay. Most antibodies used for biosensors have relatively high affinities such that the binding is not readily reversible. This does not necessarily present a problem, because in many cases sensors may be inexpensive enough to be disposable. There are other cases for which a reusable biosensor would be of value. Strategies for continuous operation or reuse of these biosensors include the kinetic measurement of analyte-induced changes in the steady-state binding of tracer to the antibody *(38)* or regeneration of the antibody (e.g., disruption of the antibody–antigen binding) using acidic buffers or chaotropic agents (*see* Chapters 3, 5, and 8).

One of the potential advantages of biosensors is their rapid operation, consequently the time required to complete the assay is another issue that must be considered. Since steady-state binding may require minutes to hours to reach (depending on the immunochemicals and assay format), many biosensor assays use rate measurements or presteady-state binding values. Because components of a biosensor system may be fabricated and stored, the time required to construct the biosensor is not usually considered of critical importance; however, the regeneration time for reusable sensors should be considered.

One area in which biosensors show a significant potential for future development is for remote, continuous, or *in situ* operation *(8)*. However, this mode of operation is a particularly challenging area for affinity-based biosensors. Because these biosensors are typically configured as competitive assays, reagents other than the analyte of interest are usually required. Several biosensor configurations based on homogeneous fluorescence energy transfer formats have been reported to accommodate this requirement. These include the use of polymers that continuously release a new supply of antibody tracer *(39)* or the use of a semipermeable membrane such that the receptor and ligand-tracer can not diffuse away from the sensor *(40,41)*.

6. Potential Applications and Future Directions

Potential application areas for affinity-based biosensors are similar to those for biosensors in general and include health care, food processing, environmental monitoring, and defense *(42)*. Although receptor and nucleic acid based biosensors show potential for future development, antibodies are at present the most highly developed recognition elements for these devices. As evidenced by the commercial success of immunoassays, primarily in the health care area but also to some extent in the food processing and environmental monitoring areas, the range of analytes, sensitivity, selectivity, and reliability afforded by these methods appears sufficient for continued use and development. The interesting question with respect to biosensors, however, is how can the direct interface of these assays to signal transduction technologies improve currently available methods as well as generate new applications?

Despite the relatively large amount of resources that have been invested by the commercial sector to develop antibody-based biosensor products, little obvious success has been realized. Nevertheless, numerous biosensor techniques have been reported, developed, and some commercialized that allow researchers to better study the kinetics, structure, and (solid/liquid) interface phenomena associated with protein-ligand binding interactions. Furthermore, with current progress in the development of versatile, inexpensive, and reliable techniques for microfabrication and protein immobilization/stabilization, affinity-based biosensors will continue to represent a significant portion of the biosensor literature and show the potential to produce commercial products in a number of areas. These biosensors will most likely find applications as stand-alone techniques, detectors in chromatographic methods, and microsensors integrated into multianalyte clinical analysis systems.

Acknowledgment

The US Environmental Protection Agency (EPA) through its Office of Research and Development (ORD) has, in part, funded the work involved in preparing this chapter. It has been subject to the Agency's peer review and has been approved for publication. The US Government has the right to retain a nonexclusive, royalty-free license in and to any copyright covering this article.

References

1. Yalow, R. S. and Berson, S. A. (1959) Assay of plasma insulin in human subjects by immunological methods. *Nature* **184,** 1648,1649.

2. Ekins, R. P. (1960) The estimation of thyroxin in human plasma by an electro-phoretic technique. *Clin. Chim. Acta* **5**, 453–459.

3. Kronick, M. N. and Little, W. A. (1974) A new immunoassay based on fluorescent excitation by internal reflection spectroscopy. *Proc. Natl. Acad. Sci. USA* **71**, 4553–4555.

4. Giaever, I. (1973) The antibody:antigen interaction: a visual observation. *J. Immunol.* **110**, 1424–1426.

5. Tromberg, B. J., Sepaniak, M. J., Alarie, J. P., Vo-Dinh, T., and Santella, R. M. (1988) Development of antibody-based fibre optic sensors for detection of benzo(a)pyrene metabolite. *Anal. Chem.* **60**, 1901–1907.

6. Turner, A. P. F., Karube, I., and Wilson, S., eds. (1987) *Biosensors: Fundamentals and Applications*, Oxford University Press, New York.

7. Marco, M.-P., Gee, S., and Hammock, B. D. (1995) Immunochemical techniques for environmental analysis: immunosensors. *Trends Anal. Chem.* **14**, 341–350.

8. Morgan, C. L., Newman, D. J., and Price, C. P. (1996) Immunosensors: technology and opportunities in laboratory medicine. *Clin. Chem.* **42**, 193–209.

9. Skladal, P. (1997) Advances in electrochemical immunosensors. *Electroanalysis* **9**, 737–744.

10. Wang, J., Cai, X., Rivas, G., Shiraishi, H., Farias, P. A. M., and Dontha, N. (1996) DNA electrochemical biosensor for the detection of short DNA sequences related to the human immunodeficiency virus. *Anal. Chem.* **68**, 2629–2634.

11. Watts, H. J., Yeung, D., and Parks, H. (1995) Real-time detection and quantification of DNA hybridization by an optical biosensor. *Anal. Chem.* **67**, 4283–4289.

12. Pandey, P. C. and Weetall, H. H. (1995) Detection of aromatic compounds based on DNA intercalation using an evanescent wave biosensor. *Anal. Chem.* **67**, 787–792.

13. Edwards, R., ed. (1996) *Immunoassays Essential Data*. Wiley, New York.

14. Nakanishi, K., Muguruma, H., and Karube, I. (1996) A novel method of immobilizing antibodies on a quartz crystal microbalance using plasma-polymerized films for immunosensors. *Anal. Chem.* **68**, 1695–1700.

15. Changeux, J. P., Devillers-Thiery, A., and Chemouille, P. (1984) Acetylcholine receptor: an alosteric protein. *Science* **225**, 1335–1345.

16. Eldefrawi, M. E. and Eldefrawi, A. T. (1973) Purification and molecular properties of the acetylcholine receptor from torpedo electroplax. *Arch. Biochem. Biophys.* **159**, 362–373.

17. Minami, H., Sugawara, M., Odashima, K., Umezawa, Y., Uto, M., Michaelis, E. K., and Kuwana, T. (1991) Ion channel sensors for glutamic acid. *Anal. Chem.* **63**, 2787–2795.

18. Sugao, N., Sugawara, M., Minami, H., Uto, M., and Umezawa, Y. (1993) Na^+/D-glucose cotransporter based on bilayer lipid membrane sensor for D-glucose. *Anal. Chem.* **65**, 363–369.

19. Taylor, R. F. (1996) Immobilization methods, in *Handbook of Chemical and Biological Sensors* (Taylor, R. F. and Schultz, J. S., eds.), IOP, Philadelphia, PA, pp. 203–219.

20. Bhatia, S. K., Shriver-Lake, L. C., Prior, K. J., Georger, J. H., Calvert, J. M., Bredhorst, R., and Ligler, F. S. (1989) User of thiol-termial silanes and hetero-bifunctional crosslinkers for immobilization of antibodies on silica surfaces. *Anal. Biochem.* **178,** 408–413.
21. Luong, J. H. T., Sochaczewski, E. P., and Guilbault, G. G. (1990) Development of a piezoimmunosensor for the detection of Salmonella typhimurium. *Ann. NY Acad. Sci.* **613,** 439–443.
22. Rogers, K. R., Kohl, S. D., Riddick, L. A., and Glass, T. R. (1997) Detection of 2,4-D using the KinExA immunoanalyzer. *The Analyst,* **122,** 1107–1111.
23. Rogers, K. R., Valdes, J. J., and Eldefrawi, M. E. (1989) Acetylcholine receptor fiber-optic evanescent fluorosensor. *Anal. Biochem.* **182,** 353–359.
24. Cornell, B. A., Braach-Maksvytis, V. L. B., King, L. G., Osman, P. D. J., Raguse, B., Wieczorek, L., and Pace, R. J. (1997) A biosensor that uses ion-channel switches. *Nature* **387,** 580–583.
25. Rabbany, S. Y., Donner, B. L., and Ligler, F. S. (1994) Optical biosensors. *Crit. Rev. Biomed. Engineer.* **22,** 307–346.
26. Jonsson, U. and Malmqvist, M. (1992) Real time biosepecific analysis. *Adv. Biosens.* **2,** 291–336.
27. Beir, F. F. and Schmidt, R. D. (1994) Real time analysis of competitive binding using grating coupler immunosensors for pesticide detection. *Biosens. Bioelectr.* **9,** 125–130.
28. Beir, F. F. and Scheller, F. W. (1996) Label-free observation of DNA-hybridiza-tion and endonuclease activity on a waveguide surface using a grating coupler. *Biosens. Bioelectr.* **11,** 669–674.
29. Guilbault, G. G. and Luong, J. H. T. (1994) Piezoelectric immunosensors and their applications in food analysis, in *Food Biosensor Analysis* (Wagner, G. and Guilbault, G. G., eds.), Marcel Dekker, New York, pp. 151–172.
30. Murastsugu, M., Fumihiko, O., Yoshihiro, M., Hosokawa, T., Kurosawa, S., Kamo, N., and Ikeda, H. (1993) Quartz crystal microbalance for the detec-tion of microgram qantities of human serum albumin: relationship between the frequency change and the mass of protein adsorbed. *Anal. Chem.* **65,** 2933–2937.
31. Starodub, N., Arenkov, P., Starodub, A., and Berezin, V. (1994) Construction and biomedical application of immunosensors based on fiber optics and enhanced chemiluminescence. *Optical Eng.* **33,** 2958–2963.
32. Danielsson, B., Mattiansson, B., and Mosbach, K. (1981) Enzyme thermistor devices and their analytical applications. *Appl. Biochem. Bioeng.* **3,** 97–143.
33. Kalab, T. and Skadal, P. (1995) A disposable amperometric immunosensor for 2,4-dichlorophenoxyacetic acid. *Anal. Chim. Acta* **304,** 361–368.
34. Sandberg, R. G., Van Houten, L. J., Schwartz, J. B. R. P., Dallas, S. M., Silva, J. C., Michael, A., and Narayanswamy, V. (1992) A conductive polymer-based immunosensor for analysis of pesticide residues. *ACS Symp. Ser.* **511,** 81.
35. Sadik, O. A. and Van Emon, J. M. (1997) Designing immunosensors for environ-mental monitoring. *Chemtech,* June, 38–46.

36. Eray, M., Dogan, N. S., Reiken, S. R., Sutisna, H., Vanwei, B. J., Koch, A. R., Moffett, D. F., Silber, M., and Davis, W. C. (1995) A highly stable and selective biosensor using modified nicotinic acetylcholine receptor (nAChR). *Biosystems* **35,** 183–188.
37. Rogers, K. R., Valdes, J. J., Menking, D., Thompson, R., and Eldefrawi, M. E. (1991) Pharmacologic specificity of an acetylcholine receptor fiber-optic biosensor. *Biosens. Bioelectron.* **6,** 507–516.
38. Anis, N. A., Eldefrawi, M. E., and Wong, R. B. (1993) Reusable fiber optic immunosensor for rapid detection of imazethapyr herbicide. *J. Agric. Chem.* **41,** 843–848.
39. Barnard, S. M. and Walt, D. R. (1991) Chemical sensors based on controlled-release polymers. *Science* **251,** 927.
40. Anderson, F. P. and Miller, W. G. (1988) Fiber optic immunochemical sensor for continuous, reversible measurement of phentoin. *Clin. Chem.* **34,** 1417–1421.
41. Meadows, D. L. and Schultz, J. S. (1993) Design, manufacture and characterization of an optical fiber glucose affinity sensor based on a homogeneous fluorescence energy transfer asssay system. *Anal. Chim. Acta* **280,** 21–30.
42. Taylor, R. F. (1996) Chemical and biological sensors: markets and commercialization, in *Handbook of Chemical and Biological Sensors* (Taylor, R. F. and Schultz, J. S., eds.), IOP, Philadelphia, PA, pp. 553–559.

2

Immunobiosensors Based on Thermistors

Kumaran Ramanathan, Masoud Khayyami, and Bengt Danielsson

1. Introduction

1.1. Calorimetric Devices

Calorimetric sensing or thermometric sensing involving immobilized biocatalysts has diversified into several areas of application since its introduction in the early 1970s. In principle, a chemical or biological process is monitored and quantified by the changes in the thermal signatures of the reacting species. From the fundamental laws in nature governing molecular reactions, virtually all reactions are associated by the absorption or evolution of heat. It was recognized quite early that most enzymatic reactions are associated with the liberation of heat. This led to the development of several generations of calorimeters that monitored biological reactions and later were successfully applied to the studies on immobilized enzymes. The different generations of calorimeters also used different approaches for heat measurement, such as isothermal, heat-conduction, and isoperibol calorimeters (*see* Chapter 13, Methods in Biotechnology, vol. 6).

1.2. The Enzyme Thermistor

In the modern isoperibol calorimeter, called enzyme thermistor (ET), the measurement is based on determination of the temperature change associated with the reactions in a microcolumn containing an immobilized biocatalyst. In a chapter in the companion volume (Chapter 13, Methods in Biotechnology, vol. 6) the application of ET to several chemical and biological problems has been extensively reviewed. In this chapter the discussion has been restricted to the application of ET to immunosensing.

From: *Methods in Biotechnology, Vol. 7: Affinity Biosensors: Techniques and Protocols*
Edited by: K. R. Rogers and A. Mulchandani © Humana Press Inc., Totowa, NJ

1.3. The TELISA

The possibility to use an ET as detector for the enzyme label in immuno-assays was earlier demonstrated by introduction of the thermometric enzyme-linked immunosorbent assay (TELISA) technique (1). To date, several variations of TELISA have been reported, as discussed below.

1.4. The Applications of TELISA

In the first TELISA assay, for the endogenous and exogenous compounds in biological fluids, the technique was used to study human serum albumin down to a concentration of 10^{-10} M (5 ng/mL). Here the normal and catalase-labeled human serum albumin compete for the binding sites on the immunosorbent column. This column contained rabbit antihuman serum albumin antibodies immobilized on Sepharose CL-4B (1).

The release of human proinsulin by genetically engineered *Escherichia coli* cells was also determined and monitored using a TELISA technique. Several M9 media samples were analyzed sequentially with the aid of a rapid auto-mated flow-though TELISA system (2). The response time for each assay was 7 min after sample injection and a single assay was complete within 13 min. Insulin concentrations in the range of 0.1–50 µg/mL could be determined by this method. This TELISA method correlated well with conventional radioim-munoassay (RIA) determinations. Reproducible performance was obtained over a period of several days even when the immobilized antibody column was stored at 30°C in the ET-unit. The sensitivity and speed of analysis were adequate for the determination of hormones, antibodies, and other biomolecules produced by fermentation.

In an alternative approach, a competitive TELISA-based assay was designed for the measurement of mouse IgGs (immunoglobulins). A heterogeneous com-petitive assay was designed for this purpose. Immobilized antibodies were packed into the column. Samples containing enzyme-labeled mouse IgGs in different concentrations were added before adding the substrate. The quantity of bound IgG was proportional to the enzyme activity detected as a large thermometric change. This flow-though system had a sample processing time of approx 6 min and offered interesting possibilities for protein monitoring during fermentation.

A TELISA based on recycling and consequent amplification of the thermal signal has also been designed. In a competitive assay mode, the sample is mixed with enzyme-labeled antigen and the concentration of bound antigen is deter-mined by the introduction of a substrate pulse. The product stream is then directed to the sampling valve of the detection column. Thereafter, the immunosorbent column is regenerated by a pulse of glycine at low pH. The whole cycle takes <13 min. The use of alkaline phosphatase as enzyme label allows amplification of the sensitivity by using phosphoenolpyruvate as substrate. A separate detection column in the ET-unit can be used for the determination of the product (pyruvate) by substrate recycling. This is accomplished by using a substrate recycling system comprising three coimmobilized enzymes: lactate dehydrogenase (LDH, which reduces pyruvate to lactate with the consumption of NADH), lactate oxidase (LOD, which oxidizes lactate to pyruvate), and catalase, which breaks down H_2O_2 (**Fig. 2**). The effect of flow rates, conjugate concentrations, and support material for the immunosorbent column was optimized. The assay had a detection limit of 0.025 μg/mL and a linear range from 0.05 to 2 μg/mL. This corresponds to a 10-fold increase in sensitivity over the unamplified system. The results demonstrated that enzymatic amplification could be employed to increase the sensitivity and reproducibility of flow injection assay-based immunosensors. The implications of these results on on-line analysis were discussed. In addition, genetically engineered enzyme conjugates have also been used in immunoassays. Thus, a human proinsulin-*E. coli* alkaline phosphatase conjugate was used for the determination of insulin and proinsulin *(3)*.

A sandwich-type *(4)* flow-injection binding assay for quantifying IgGs from various sources has also been developed. The assay is based on a pseudo-immunological reaction between protein A from *Staphylococcus aureus* and immunoglobulin G from different species. Protein A was immobilized on a solid support and a fusion protein of protein A and β-galactosidase from *E. coli* was used for detection. The fusion protein is produced with a temperature-inducible recombinant *E. coli* strain. A sandwich structure was formed by the subsequent injection of IgG and fusion protein into the buffer stream, flowing though the immobilized protein A column. The amount of enzyme activity bound was proportional to the amount of IgG bound and was measured by pumping a lactose solution as substrate for β-galactosidase through the protein A column. Lactose is converted to glucose and galactose. The detector was an ET that measures the heat evolved in the enzymatic conversion of glucose by coimmobilized glucose oxidase and catalase. The assay took about 16 min at a flow rate of 0.6 mL/min with a lower detection limit of 33 pmol per injection of rabbit IgG. The precision of replicate measurements had a standard deviation of 4–5%, and the column could be used for more than 50 cycles *(6,7)*.

1.5. The Construction of ET

The enzyme thermistor (ET) could be designed using Plexiglas where a water bath controls the temperature, or modified by using metal blocks for temperature control (*see* Fig. 1 in Chapter 13, Methods in Biotechnology, vol. 6). Essentially the ET consists of a plastic column, which can hold varying amounts of the enzyme/antibody depending on the volume (<1 mL) of the column and the nature of the adsorbent. The inlet solution (either the

circulating buffer or the analyte (to be measured) passes though acid-proof
steel tubing that acts as an efficient heat exchanger. The solution (containing
the analyte) then flows into the column and reacts with the enzyme/antibody
immobilized on the column, resulting in a temperature change. The latter is
measured at the top of the column with a thermistor attached to a short gold
tube at the exit of the column. The signal is in the form of an unbalance
signal of a sensitive Wheatstone bridge using a reference thermistor at a
stable temperature of the calorimeter. The column and the thermistor probe
are inserted into the cylindrical aluminum calorimeter, enveloped by poly-
urethane foam for thermal insulation, enabling more sensitive measurements
(*see* **refs. 8–10**).

2. Materials
2.1. Purification of Insulin Antibodies

1. Antiporcine insulin serum from Miles (Rehovot, Israel), beef insulin from
   Sigma (St. Louis, MO), Sepharose 4B from Pharmacia (Uppsala, Sweden),
   2,2,2-trifluoroethene sulfonyl chloride from Fluka (Buchs, Switzerland) and
   acetone (HPLC grade) from Sigma.
2. 50 mM Sodium phosphate buffer containing 0.15 M NaCl at pH 7.4.
3. 0.5 M NaHCO$_3$ and 0.2 M NaHCO$_3$ prepared in deionized water (18 MΩ
   resistance).

2.2 Synthesis and Purification
of the Insulin Peroxidase Conjugate

1. Horseradish peroxidase type-VI (EC 1.11.1.7), activity 250–330 U/mg solid from
   Sigma.
2. 21.4 mg/mL sodium metaperiodate in deionized water.
3. Dialysis medium: 1 mM sodium acetate buffer, pH 4.4.
4. Dialysis medium: 10 mM sodium bicarbonate buffer, pH 9.5.
5. Dialysis medium: 0.2 mM sodium bicarbonate buffer, pH 9.5.
6. Dialysis medium: 10 mM sodium bicarbonate buffer, pH 9.5, containing 0.5 mg/mL
   NaBH$_4$.
7. 0.1 M PBS containing 0.15 M NaCl at pH 7.4.

2.3. Testing of Samples with TELISA

1. Substrate for peroxidase: 2 mM H$_2$O$_2$, 14 mM phenol, and 0.8 mM 4-amino-
   antipyrine in PBS.
2. Washing buffer: 0.2 M glycine-HCl buffer, pH 2.2.

2.4. Determination of Alkaline Phosphatase Activity

1. Assay buffer: 1 mM p-nitrophenyl phosphate, 1 mM MgCl$_2$, and 0.1 M glycine at
   pH 10.4.

2.5. TELISA Amplification Assay

1. Enzymes: Catalase (EC 1.11.1.6) from beef liver, 65000 U/mg protein, lactate dehydrogenase (EC 1.1.1.27) from hog muscle, 550 U/mg protein, lactate oxidase (EC 1.1.3.2) from *Pediococcus species*, 20 U/g mg protein and alkaline phosphatase (EC 3.1.3.1) from calf intestine, 3000 U/mg protein.
2. Immunosorbent column: Eupergit C from Rohm Pharma, Weiterstadt, Germany. Sephacryl S-200, Sepharose, 4B, DEAE-Sepharose, and tresyl chloride from Pharmacia, Sweden.
3. Coupling buffer: 1 M potassium phosphate, pH 7.5, containing 7 mM p-hydroxy benzoic acid ethyl ester.
4. Washing buffer: 0.1 M potassium phosphate, pH 7.5.
5. Antigen binding buffer: 0.1 M Tris-HCl, pH 7.4, containing 0.15 M NaCl.
6. Blocking buffer: 5 g% BSA, 0.05% Tween-20, 0.5 M NaCl, 0.1 M Tris-HCl, pH 7.4 (1X) or 10 g% BSA, 0.1% Tween-20, 1 M NaCl, 0.2 M Tris-HCl, pH 7.4 (2X).
7. Substrate assay buffer: 0.15 M NaCl, 0.02 M Tris-HCl, pH 10.0.
8. Substrate solution: 1.5 mM phosphoenol pyruvate (1 mL), 0.15 M NaCl, 0.02 M Tris-HCl, pH 10.0.
9. Recycling buffer: 1 mM NADH, 0.14 M NaCl, 0.09 M sodium phosphate, 0.018 M Tris-HCl, pH 7.0.
10. Regeneration buffer: 0.2 M glycine-HCl, pH 2.2.

3. Methods

The TELISA technique is demonstrated below for assay of proinsulin using a competitive enzyme immunoassay and a flow-injection and enzyme-amplification-based thermometric ELISA.

3.1. Assay for Proinsulin Using a Competitive Enzyme Immunoassay

3.1.1. Purification of the Insulin Antibodies

3.1.1.1. ACTIVATION OF IMMUNOSORBENT COLUMN (SEPHAROSE 4B) USING TRESYL CHLORIDE

1. Wash Sepharose 4B (100 mL settled gel) with 1 L water in a sintered glass filter funnel. Successively wash the gel with 1 L each of 30:70, 60:40, and 80:20 acetone:water mixtures. Wash twice with acetone and finally three times with dry acetone.
2. Place the gel in a dried beaker containing 100 mL dry acetone and 10 mL dry pyridine (pyridine is to neutralize the liberated HCl during activation). While stirring, add dropwise 2 mL tresyl chloride over a period of 1 min.
3. Continue the stirring for 10 min at room temperature.
4. Wash the activated gel twice with 1 L each of acetone, 30:70, 50:50, and 70:30 of 1 mM HCl:acetone, and finally with 1 mM HCl. Tresylated Sepharose 4B can be stored for several weeks at 4°C in 100 mL, 1 mM HCl without losing coupling efficiency.

3.1.1.2. Coupling of Insulin to Activated Sepharose 4 B (*see* **Note 7**)

1. Suspend 1 g of tresyl chloride-activated matrix in 10 mL of 0.2 M sodium phosphate buffer, pH 7.5., in which 5 mg of beef insulin is added.
2. Stir the gel suspension at 4°C using a paddle stirrer for 24 h.
3. Wash the coupled gel extensively with 0.2 M sodium phosphate buffer, pH 7.5, containing 1.0 M NaCl and water to remove inactivated ligand.
4. Block the excess groups on the gel by suspending the gel in 100 mL of 1.0 M ethanolamine, pH 9.0, and stirring for 1 h at room temperature.
5. Finally, wash the gel extensively with 1.0 M NaCl and water.

3.1.1.3. Isolation and Purification of Anti-Insulin Antibodies (*see* **Note 1**)

1. Apply antiporcine insulin serum to 2.5 g of wet gel packed in the column.
2. Elute the antibodies from the column under reverse flow with 50 mM sodium phosphate buffer containing 0.15 M NaCl at pH 7.4 until there is no absorbance at 280 nm.
3. Mix the eluted peak immediately with 1.5 mL of 0.5 M NaHCO$_3$ and dialyze against 0.2 M NaHCO$_3$ for 24 h at 4°C.
4. Couple the purified antibodies 0.2 mg/mL to Sepharose 4B (1 mL/g wet gel) activated with 6 µL tresyl chloride/g of gel.

3.2. Synthesis and Purification
of the Insulin-Peroxidase (HP) Conjugate (see Note 2)

1. Dissolve 4 mg/mL of HP in 4 mL of deionized water.
2. Add 0.8 mL of freshly prepared sodium metaperiodate to the HP solution and stir for 30 min at 27°C.
3. Dialyze the peroxidase-periodate solution against sodium acetate buffer at 4°C overnight.
4. Add 8 mg beef insulin in 2 mL HCl (50 mM) and 2 mL NaOH (50 mM) and dialyze against sodium bicarbonate buffer (10 mM), at 4°C overnight (*see* **Note 9**).
5. Adjust the pH of the peroxidase-aldehyde dialyzed solution (*see* **step 3**) between 9 and 9.5 by addition of 50–200 µL of 0.2 M sodium bicarbonate buffer.
6. Mix the insulin solution with the peroxidase-aldehyde solution and stir at 27°C for 2 h.
7. Dialyze the insulin-peroxidase solution (*see* **step 6**) against sodium bicarbonate buffer (10 mM) containing NaBH$_4$ for 2 h at 4°C.
8. Dialyze the insulin-peroxidase conjugate (*see* **step 7**) overnight at 4°C against PBS.
9. Chromatograph the conjugate using S-200 (35 × 2.5 cm column) using PBS and collect 3 mL fractions at a flow rate of 0.25 mL/min.
10. Measure absorbance at 403 and 280 nm and store the fractions containing the conjugate peak at 4°C until used.

3.3. Procedure for Automatic Sampling

1. Arrange the experimental set up as shown in Fig. 1 in **ref. 2**.
2. Set the injection and cycle timers at 2 min each.

3. Activate the autosampler to draw samples from the first position on the sampler rack and fill the injection loop with 0.5 mL of sample with the help of a peristaltic pump operating at approx 2 mL/min.
4. Wait for 20 s and activate the injection timer, to inject the sample into the thermistor column.
5. Set the speed of the peristaltic pump circulating the PBS buffer into the thermistor column (containing anti-insulin antibody from **step 4** in **Subheading 3.1.1.3.**) at 0.6 mL/min. After 2 min the valve returns to the load position for refilling the sample.
6. The enzyme thermistor must be maintained at a constant temperature of 30°C.
7. During the subsequent 2 min the thermistor column gets washed with the PBS buffer while the signal returns to the background value.

3.4. The Automated Competitive TELISA

1. On the sampling rack the first test tube contains 0.5 mL cell-free fermentation sample M9 media plus 0.5 mL insulin-peroxidase conjugate. The stock of the conjugate is diluted 1:20 in PBS (*see* **Note 9**).
2. Fill the second tube with 1 mL substrate for peroxidase (*see* **Subheading 2.3., step 1**).
3. Measure the heat of reaction between the substrate and peroxidase (**Fig. 1A**).
4. Following the measurement inject 1.0 mL glycine-HCl buffer from a third test tube. This washes away the insulin-peroxidase conjugate and proinsulin from the immunosorbent and refreshes the column for detecting the next sample.
5. The column can be stored in the ET without significant loss in enzyme activity (**Fig. 1B**) (*see* **Note 3**).

3.5. A Flow-Injected Thermometric ELISA Using Enzyme Amplification

3.5.1. Synthesis and Purification of the Proinsulin-Alkaline Phosphatase Conjugate

The proinsulin-alkaline phosphatase conjugate is prepared in a similar fashion to the procedure described in **Subheading 3.2.** for synthesis of insulin-peroxidase conjugate. The activity of the proinsulin-alkaline phosphatase conjugate is tested as follows.

1. Add 0.1 mL of the conjugate to 1.9 mL of the assay buffer (*see* **Subheading 2.4., item 1**).
2. Incubate for 1 h and record the absorbance at 405 nm.

3.5.2. Preparation of the Immunosorbent Column for Amplification Assay

1. Dialyze 0.9 mg of affinity purified anti-insulin antibody (**Subheading 3.1.1.3.**) against coupling buffer.

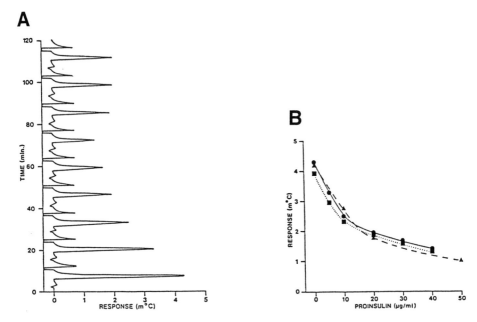

Fig. 1. **(A)** The thermometric response peaks as plotted on the recorder. In ascending order, a standard curve of porcine proinsulin in the concentration range 40, 30, 20, 10, 5, and 0 µg/mL followed by a triple injection of the fermentation supernatant of *E. coli* producing human proinsulin. The disturbance in the baseline following each temperature peak is owing to the solvation effects from the glycine wash of the column, but does not affect the measurement. The detailed procedure is described in **Subheading 3.1. (B)** The variation in the thermometric standard curve with immobilized anti-insulin antibodies and the column stored in the enzyme thermistor (△) d 1, (□) d 2, and (●) d 5. The conjugate was diluted 1:10 with phosphate-buffered saline (PBS) (Reproduced from **ref. 2** with permission).

2. Adjust the final volume to 1 mL and mix with 0.25 g of Eupergit C.
3. Allow the coupling to proceed for 72 h at 27°C.
4. Wash the column with 25 mL of 0.1 *M* washing buffer, pH 7.5, at least five times prior to packing in 20 × 7 mm glass column.

3.5.3. Preparation of the Recycling Column for Amplification Assay

Prepare the substrate recycling column by mixing LDH 1100 U, LOD 36 U, and CAT 130000 U in 1 mL of phosphate buffer pH 7.5 and following a similar procedure as described in **Subheading 3.1.1.2., step 2**.

Assay procedure for the TELISA amplification assay 1 (*see* **Notes 4, 5,** and **7**):

1. Place the immunosorbent column in the enzyme thermistor unit maintained at 37°C.

2. Equilibrate with antigen binding buffer at a flow rate of 0.5 mL/min and inject 1 mL of the blocking buffer (1X).
3. Dilute the sample/conjugate mixture (0.1 mL) with an equal volume of blocking buffer (2X).
4. Equilibrate the column with substrate assay buffer for 5 min and inject the substrate solution.
5. Collect the effluent from the immunoabsorbent column containing the pyruvate (empirically determined).
6. Mix the solution (**step 5**) with 1 M sodium phosphate at pH 6.3.
7. Mount the recycling column in a second thermistor at 30°C.
8. Equilibrate the recycling column with recycling buffer.
9. Inject the sample (as in **step 5**) and maintain a flow rate of the buffer at 1 mL/min.
10. Regenerate the first column, i.e., the immunosorbent column, using the regeneration buffer and then equilibrate in binding buffer for 3 min before the next sample injection.

4. Notes

1. From 1 mL of antiporcine insulin guinea pig serum about 0.6 mg of antibodies can be obtained.
2. The enzyme-antigen ratio can be calculated by comparing the A_{403} and A_{280} values, which are quoted as 22 and 7, respectively, for a 1% (w/v) solution of HP. However, these values can be significantly lower, i.e., A_{403} and A_{280} could be 5 and 5, respectively, for a 1% solution. However, the A_{280} for insulin is approx 10. Based on these values, the molar ratio of the stock conjugate solution can be calculated to be 1:2.5 containing 0.48 mg/mL HP and 0.17 mg/mL insulin.
3. The immunosorbent column could be stored at 4°C in PBS containing 0.05% NaN_3. In this case, the column has to be equilibrated at least for 2 h prior to further experiments. However, if the column is stored at 25°C in the enzyme thermistor for 1 wk with continuous circulation of PBS containing 0.05% NaN_3 then the column can be reused immediately.
4. In order to obtain reproducible and consistent results, it is essential to maintain a constant temperature, pH, injection volume, time, and flow rate in the various batch of experiments, especially in the flow injection analysis mode.
5. Any change in the sample volume or the volume of the immobilized column can result in diminished absolute peak height and has to be compensated by increasing the conjugate concentration or increasing the immobilized antibody concentration. If the flow rate is increased to minimize the assay time, the absolute thermometric peak height would diminish and needs to be offset. For Sepharose 4B, the maximum tolerance in flow rate is 11.5 and 26 mL/cm^2/h for Sepharose CL-4B.
6. In the case of the amplification assay, it is possible to increase the sensitivity and reproducibility by optimizing the conjugate concentration. Reducing the conjugate concentration below normal detection limits allows the bound enzyme (alkaline phosphatase) to perform reproducibly within the linear range. This results in an increase in the linear range and the reproducibility of the assay. In

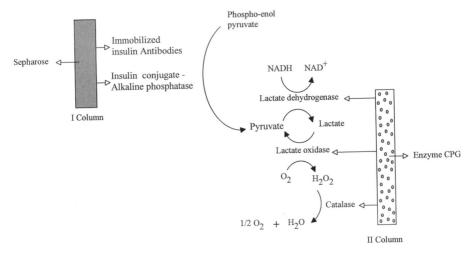

Fig. 2. A schematic of TELISA coupled to detection based on substrate recycling using an enzyme sequence. Column I consists of immobilized anti-insulin antibody covalently coupled to sepharose. Column II consists of lactate dehydrogenase, lactate oxidase, and catalase, immobilized on amino-silanized CPG (Controlled Pore Glass) by crosslinking with glutaraldehyde. Phosphoenol pyruvate is the substrate for the alkaline phosphatase. The pyruvate produced in column I is detected by column II.

         addition, use of recombinant conjugate (*see* **ref. 6**) is an advantage compared to the use of organically synthesized conjugates.

7. Tresyl chloride activation should be carried out in a well ventilated hood. Use only acetone during activation to prevent hydrolysis of tresyl chloride.

8. In the above assay two columns are employed as depicted in **Fig. 2**. The first column consists of immobilized antiporcine insulin antibodies that bind to labeled insulin or unlabeled insulin, competitively. The label is the enzyme alkaline phosphatase. The second column consists of coimmobilized LDH, LOD, and catalase for the amplification. The heat generated in the second column is sensed thermometrically.

9a. The M9 media is used for the fermentative production of proinsulin by *E. coli* EC 703 from Biogen S.A. (Geneva, Switzerland). The bacteria releases the proinsulin molecule into the media by cell lysis.

b. In such an assay there is a competition between proinsulin from the fermentation sample and the insulin-peroxidase conjugate to bind to the anti-insulin antibody column.

References

1. Mattiasson, B., Borrebaeck, C., Sanfridson, B., and Mosbach, K. (1977) Thermometric enzyme linked immunosorbent assay: TELISA. *Biochim. Biophys. Acta.* **483,** 221–227.

2. Birnbaum, S., Bülow, L., Hardy, K., Danielsson, B., and Mosbach, K. (1986) Automated thermometric enzyme immunoassay of human proinsulin produced by *Escherichia coli. Anal. Biochem.* **158**, 12–19.
3. Mecklenburg, M., Lindbladh, C., Li, H., Mosbach, K., and Danielsson, B. (1993) Enzymatic amplification of a flow-injected thermometric enzyme-linked immunoassay for human insulin. *Anal. Biochem.* **212**, 388–393.
4. Scheper, T., Brandes, W., Maschke, H., Plötz, F., and Müller, C. (1993) Two FIA-based biosensor systems studied for bioprocess monitoring. *J. Biotech.* **31**, 345–356.
5. Brandes, W., Maschke, H.-E., and Scheper, T. (1993) Specific flow injection sandwich binding assay for IgG using protein A and a fusion protein. *Anal. Chem.* **65**, 3368–3371.
6. Lindbladh, C., Persson, M., Bülow, L., Stahl, S., and Mosbach, K. (1987) The design of a simple competitive ELISA using human proinsulin-alkaline phosphatase conjugates prepared by gene fusion. *Biochem. Biophys. Res. Commun.* **149(2)**, 607–614.
7. Scheper, T., Brandes, W., Grau, C., Hundeck, H. G., Reinhardt, B., Ruther, F., Plotz, F., Schelp, C., and Schügerl, K. (1991) Applications of biosensor systems for bioprocess monitoring. *Anal. Chim. Acta.* **249**, 25–34.
8. Hundeck, H.-G., Sauerbrei, A., Hubner, U., Scheper, T., and Schügerl, K. (1990) Four-channel enzyme thermistor system for process monitoring and control in biotechnology. *Anal. Chim. Acta.* **238**, 211–221.
9. Scheper, T. H., Hilmer, J. M., Lammers, F., Müller, C., and Reinecke, M. (1996) Review: biosensors in bioprocess monitoring. *J. Chomatog. A.* **725**, 3–12.
10. Danielsson, B. and Mattiasson, B. (1996) Thermistor-based biosensors, in *Handbook of Chemical & Biological Sensors* (Taylor, R. F. and Schultz, J. S., eds.), Institute of Physics Publishing, Philadelphia, pp. 1–17.
11. Borrebaeck, C., Mattiasson, B., and Svensson, K. (1978) A rapid non-equilibrium enzyme immunoassay for determining gentamycin, in *Enzyme Labelled Immunoassay of Hormones and Drugs* (Pal, S. B., ed.), Gruyter, Berlin, Germany, pp. 15–28.
12. Wilson, M. B. and Nakane, P. K. (1978) in *Immunofluorescence and Related Staining Techniques* (Knapp, W., ed.), Elsevier/North Holland Biomedical, Amsterdam, The Netherlands, pp. 215–224.

3

Affinity Biosensing Based on Surface Plasmon Resonance Detection

Bo Liedberg and Knut Johansen

1. Introduction

Interaction between molecules is the basis for life in cells and higher organisms. Detailed investigations of the intimate relationship between the structural properties of interacting molecules and their biological function, and the biological consequences of their interactions are therefore central research topics in modern molecular biology. The molecular interactions referred to here can be of many types; for example, antibody–antigen, receptor–ligand, and DNA–protein. The above issues must generally be addressed on different levels of molecular complexity, including organisms, cells, proteins, and peptides, as well as low-mol-wt species like hormones, vitamins, amino acids, sugars, and so forth, by employing a broad range of biochemically and biologically oriented methods and assays. Development of new technology with improved sensitivity, specificity, and speed is therefore exceedingly important in order to be able to meet the increasing demands from the molecular biologists.

We describe in this chapter a label-free optical method capable of monitoring biological interaction phenomena at surfaces in real time, i.e., under continuous flow conditions, in which one of the molecules in the so-called interaction pair, the ligand, is covalently attached to the surface *(1–4)*. The optical-detection method is based on total internal reflection (TIR) and surface plasmon resonance (SPR) *(5–7)*, a collective oscillation of the electrons with respect to the nuclei in the near surface region of certain metals like gold, silver, and aluminum. The surface plasmon oscillation can be regarded as an optical wave that is driven by an external light source. It is a strongly localized wave that propagates along the interface between the metal and the ambient medium (e.g., a buffer or a biofluid). This wave is extremely sensitive to

From: *Methods in Biotechnology, Vol. 7: Affinity Biosensors: Techniques and Protocols*
Edited by: K. R. Rogers and A. Mulchandani © Humana Press Inc., Totowa, NJ

changes in refractive index near the metal surface, for example, caused by adsorption or binding of biomolecules to the surface. The sensing signal in an SPR experiment is defined as the change in angle of incidence Θ_{sp} of an incident light beam, and is in this chapter denoted as $\Delta\Theta_{sp}$. It is, for small angular shifts, proportional to the changes in refractive index and consequently to the mass concentration of the biomolecules at the surface of the metal.

The outcome of an SPR experiment is strongly dependent on the accessibility and presentation of the active region's "epitopes" of the immobilized ligand. In this chapter, we describe two reliable methods for covalent attachment of ligands to the sensing surface. We also describe a series of experiments concerning the structure–function relationship of a recombinant human granulocyte-macrophage colony-stimulating factor to demonstrate the applicability of SPR for affinity biosensing.

1.1. Information Obtainable with the SPR Biosensor

The interest in surface plasmon resonance has grown dramatically since 1990 when it was first introduced as the detection principle in a biosensor system for real-time biospecific-interaction analysis (BIA) (1–3). It is a fast, sensitive, and reliable method that can be used to answer the following fundamental questions about the interacting molecules:

1. How many are there? Concentration determination.
2. How fast and strong is the interaction? Determination of association and dissociation rate constants and affinity.
3. How do the molecules interact? Determination of active binding regions and relative binding patterns, e.g., "epitope mapping."
4. How specific is the interaction? Molecule or class of molecules.

1.1.1. Basic Theory Behind Surface Plasmon Detection

The most commonly used method for setting up a surface plasmon at a metal-ambient interface is schematically outlined in **Fig. 1**. This setup is often referred to as the Kretschmann configuration (8), and it is based on total internal reflection in a glass prism onto which a thin metal film is deposited. For thin metal films, with thicknesses much less than the wavelength of the light ($\lambda \approx 400$–800 nm in the visible), total internal reflection occurs at an angle of incidence $\Theta \geq \Theta_c =$ arcsin $\sqrt{\varepsilon_a}/\sqrt{\varepsilon_g}$), where ε_a and ε_g are the dielectric functions of the ambient and glass prism, respectively. For $\Theta \geq \Theta_c$ the reflected light intensity approaches unity, and no propagating light beam is refracted into the ambient medium. However, a part of the light, the so-called evanescent field, penetrates outside the glass. If the metal deposited on the prism is sufficiently thin, $d \approx 50$ nm, then the evanescent field can penetrate through the metal and set up a surface plasmon at the metal-ambient interface. A surface plasmon is, as mentioned

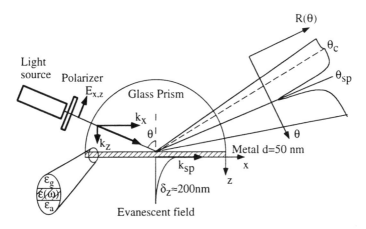

Fig. 1. Schematic illustration of the Kretschmann configuration used for optical excitation of surface plasmons. In most cases the light source consists of a light-emitting diod (LED) or a laser. Total internal reflection occurs for $\Theta \geq \Theta_c$ = arcsin $(\sqrt{\varepsilon_a}/\sqrt{\varepsilon_g})$, provided that $\sqrt{\varepsilon_g} > \sqrt{\varepsilon_a}$. The SPR phenomenon can be observed by varying Θ until $k_x = k_{sp}$. When Θ approaches Θ_{sp} the reflected light intensity $R(\Theta)$ drops dramatically. The depth, width, and shape of the $R(\Theta)$ curve depend on the thickness of the metal and on the optical constants of the prism, metal, and ambient medium, respectively. The polarizer allows only the active p-polarized component of the incident light to be reflected at the interface. Also shown is the evanescent electric field that extends a few hundred nanometers into the ambient medium. It defines the maximum thickness (interaction volume) within which the SPR phenomenon is sensitive to changes in the optical constants of the ambient medium.

above, a wave that propagates along the interface between the metal and the ambient medium. The propagation vector k_{sp} for such a surface-bound wave *(6–8)* (**Fig. 1**), can be written as follows:

$$k_{sp} = \frac{\omega}{c} \cdot \sqrt{\frac{\varepsilon(\omega) \cdot \varepsilon_a}{\varepsilon(\omega) + \varepsilon_a}} \qquad (1)$$

where ω is the frequency of the light $(2\pi c/\lambda)$, $\varepsilon(\omega)$ the dielectric function of the metal, and c the speed of light in a vacuum.

The actual SPR experiment is to tune the propagation vector of the incident light in the prism $k = \sqrt{\varepsilon_g} \cdot \omega/c$, or more precisely, the surface-parallel component of the propagation vector of the incident light, $k_x = k \cdot \sin\Theta$, for a given ω until it coincides with k_{sp}. Thus, by simply varying the angle of incidence of the light beam between $\Theta_c > \Theta > 90°$, one can find the resonance condition at which k_x equals k_{sp}, **Eq. 2**. At this particular k_x-value (or angle of incidence $\Theta = \Theta_{sp}$) most of the incident light is transferred into the surface plasmon wave and a sharp minimum in the reflected light intensity (the resonance $R(\Theta)$ curve)

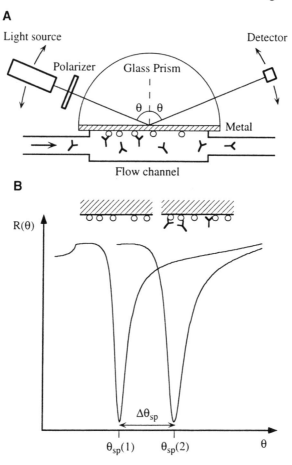

Fig. 2. Comparison between scanning **(A–B)** and fan-shaped **(C–D)** optics used in SPR detection. The two $R(\Theta)$ curves in (B) obtained before and after the biomolecular interaction are used to determine the SPR signal, e.g., $\Delta\Theta_{sp} = \Theta_{sp}(2) - \Theta_{sp}(1)$.

is observed. The surface plasmon can be described as a transverse magnetic (TM) wave phenomenon and can thus only be observed for p-polarized light *(5,6)*, where p stands for parallel to the plane of incidence.

$$\sqrt{\varepsilon_g} \cdot \sin\Theta_{sp} = \sqrt{\frac{\varepsilon(\omega) \cdot \varepsilon_a}{\varepsilon(\omega) + \varepsilon_a}} \qquad (2)$$

The prism material, light source, and metal are normally not changed during an SPR experiment. The only way to influence the resonance condition, **Eq. 2**, using monochromatic light is to manipulate the dielectric function ε_a of the ambient medium. Rewriting **Eq. 2** by assuming that $|\varepsilon(\omega)| >> \varepsilon_a$ yields that

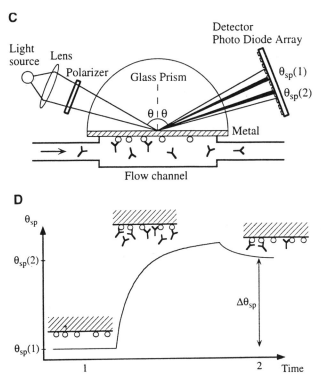

Fig. 2. (*continued*) The fan-shaped optics in (**C**) allows a predefined range of angles of incidence to be investigated simultaneously. The change in angle of incidence is followed in real-time to obtain the so-called sensorgram $\Theta_{sp}(t)$ (D).

$\sin\Theta_{sp} \propto \varepsilon_a$. Utilizing the optical relation $\sqrt{\varepsilon} = n$, where n is the refractive index, gives that $\sin\Theta_{sp} \propto n_a$. Thus, a small shift in the resonance angle $\Delta\Theta_{sp}$ (the resonance or SPR signal) is, for a given experimental setup (light source, prism, metal, buffer), proportional to the local change in refractive index Δn_a, for example, caused by interaction between biomolecules on the surface of the metal. The change in refractive index Δn_a can furthermore be used to determine the surface concentration Γ of biomolecules through the de Feijter formula *(9)*.

$$\Gamma = \frac{d \times \Delta n_a}{\delta n / \delta c} \tag{3}$$

where $\delta n / \delta c$ is a constant for the biomolecule (≈ 0.2 cm^3/g for many proteins), and d is the thickness of the biomolecular film, or the thickness of the interaction volume within which the interaction occurs (*see* **Subheading 1.2.**).

Figure 2A illustrates the principles of SPR detection of a biomolecular interaction. The sensor surface consists typically of a thin layer of gold onto which the ligand has been immobilized (e.g., an antigen). The first step in an SPR experiment is to record the background $R(\Theta)$-curve for the immobilized antigen, before the interaction, under continuous flow of the pure buffer in the flow channel, $\Theta_{sp}(1)$ in **Fig. 2B**. An analyte pulse containing an antibody directed against the immobilized antigen is then injected over the sensor surface, and an antigen–antibody complex is formed. A new $R(\Theta)$ curve is recorded after a given contact time with the antibody solution and after rinsing with buffer $\Theta_{sp}(2)$. The SPR signal, e.g., as a result of the formation of a stable antigen–antibody complex, can now be determined as the change in Θ_{sp} between the second and first minima $\Delta\Theta_{sp} = \Theta_{sp}(2) - \Theta_{sp}(1)$, as shown in **Fig. 2B**. However, the setup in **Fig. 2A** displays a number of serious limitations. The most important limitation is related to the fact that one has to scan both the source and detector with respect to the prism to obtain the $R(\Theta)$ curve. This scanning procedure is normally quite time-consuming, making it difficult to study fast association and dissociation phenomena between the interacting molecules. One way to overcome this problem is to employ optics, without any movable parts. **Fig. 2C** schematically shows an experimental setup based on so-called fan-shaped optics *(10)*. The "scanning" light source in **Fig. 2A** is replaced by a focused beam, which, within certain limits, provides a continuum of angles of incidence. The single-element detector is replaced by a photodiode array. The angle of incidence at which resonance occurs, represented by dark lines in **Fig. 2C**, can be observed as a minimum of the reflected intensity for a given pixel on the photo diode array. The short readout time for photodiode arrays enables us to follow the resonance minimum Θ_{sp} in real-time. Thus, information about the kinetics of the interaction can be obtained with this setup by following the change in resonance angle Θ_{sp} with time, the so-called sensorgram, **Fig. 2D**.

1.2. Design of Sensing Interfaces: Physical and Chemical Considerations

The surface plasmon wave is strongly localized to the interface. The depth sensitivity ($e^{-z/\delta z}$) of the surface plasmon is limited by the penetration depth δ_z of the evanescent field into the ambient medium. Typical values of δ_z in the visible region of the spectrum are 200–300 nm. Thus, molecular events (interactions) within the first few hundred nanometers can be detected by SPR, whereas those occurring far away from the interface cannot. An interesting question now appears. How can one use the extension of the evanescent field in the most efficient way? Experiments and simulations of the SPR phenomenon *(11–13)* show that the biomolecular interaction preferably should occur in a flexible coupling matrix extending about

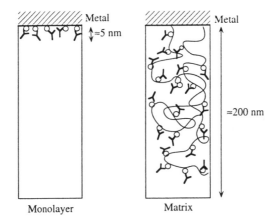

Fig. 3. Monolayer (2D) vs matrix (3D) interaction analysis. The arrows define the interaction volumes, which are about 5 and 200 nm thick, respectively.

100–200 nm away from the metal surface. More biomolecules can bind their counterparts per surface area in such a three-dimensional (3D) matrix than in a 2D monolayer array (**Fig. 3**). The loading capacity and the dynamic range of the biosensor system is thereby increased by using an extended 3D-coupling matrix. Radiolabeling experiments reveal that the capacity of a carboxymethylated matrix, about 200-nm thick, is approx 50 ng/mm^2 *(11)*, which represents about 10–15 densely packed monolayers of a small to medium-sized 20 kDa protein with a diameter of about 3.5 nm. The extension of the matrix also defines the thickness d of the interaction volume referred to in **Eq. 3, Subheading 1.1.1.**

Steric hindrance and blocking of epitopes can dramatically reduce the response of a densely packed array of recognition centers (ligands) in a monolayer, favoring again the matrix approach in which the ligands are physically separated from each other and thus less prone to influence the binding to its nearest neighbor. A few important chemical and biochemical issues that should be considered in the design of biosensing matrices are:

1. The matrix should be flexible enough to allow the interaction between the molecules to occur in a natural "solution-like" environment.
2. The matrix should be selected to have a low nonspecific binding of biomolecules and low activation levels of biologic cascade reactions. This is extremely important in applications with complex biofluids like serum or plasma.
3. The matrix should provide an attractive chemical handle for the immobilization of biomolecules (ligands, antigens, DNA, and so forth) using reliable organic synthetic chemistry.

1.3. Immobilization Strategies and Liquid Handling

The experiments presented in this chapter are all performed using a BIAcore biosensor *(1–3)* a commercial system from Pharmacia Biosensor AB* (Uppsala, Sweden). The sensing chip in this system consists of a glass slide onto which a thin gold film has been deposited. A linker layer of OH-terminated alkanethiols is assembled on the gold surface for covalent attachment of the sensing matrix, which consists of carboxymethylated dextran chains *(14,15)*. The carboxyl(carboxylate) groups, approximately one per glucose unit, act as anchoring positions for the ligand to be immobilized. They are also used to electrostatically attract positively charged ligands (proteins) in solutions of low ionic strength during the immobilization sequence *(15)*.

The general philosophy adopted in BIA involves a minimum of derivatization of the ligand before immobilization. Therefore, a series of synthetic protocols have been developed for covalent immobilization of ligands, which all are based on nucleophilic displacement of reactive *N*-hydroxysuccinimide esters attached to the COOH groups in the dextran chains *(15)*. These *N*-hydroxy-succinimide ester groups react primarily with nucleophilic groups on the native biomolecule, such as the N-terminal α-amino group, the ϵ-amino group of lysine, and the thiol group of cysteine. In this chapter, we describe two immobilization methods: amine coupling and thiol-disulfide exchange coupling. The latter will be demonstrated along with the application examples in **Subheading 2.2.** Accurate and reproducible immobilization of the ligands also requires an efficient liquid-handling system for transportation of the biomolecules (ligands, analyte) to the sensing surface.

1.3.1. Liquid Handling

Because SPR is a surface-sensitive technique, it is important to have an efficient mass transport, i.e., to allow the analyte molecules to diffuse to the surface. It is also important to have small dead-space volumes to obtain sharp analyte pulses. The microfluidic system in the biosensor system fulfills these objectives. It is an integrated fluid cartridge (IFC), made of plastic and silicone rubber *(16)*. The IFC is a flow-injection analysis (FIA) system consisting of a sample loop of approx 100 μL that is filled by the autosampler. There is normally a constant flow of buffer over the SPR surface to obtain a stable baseline. During measurement the analyte solution is automatically switched over the SPR surface by air-controlled microvalves in the silicone rubber cartridge. The flow rate is controlled by an external syringe pump. The valves are computer controlled for obtaining accurate sample volumes and contact times.

---

*Pharmacia Biosensor AB has changed its name to Biacore AB.

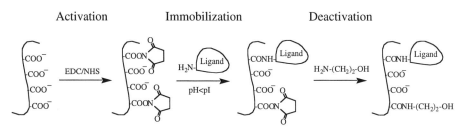

Fig. 4. Illustration of the activation–immobilization–deactivation sequence used in the amine coupling kit.

1.3.2. Amine Coupling

The major steps in the activation–immobilization–deactivation coupling sequence are outlined in **Fig. 4.** The sequence relies on the presence of accessible amine groups on the ligand. This requirement is normally fulfilled for a broad range of proteins. The conditions (protein concentration, pH, ionic strength, and such, of the protein solution) given below for the amine coupling sequence should only be regarded as representative values. In cases when extremely accurate and reproducible immobilization levels are required, for example, in an assay for concentration determination of a specific analyte, each of the above parameters must be optimized individually. A detailed description of the optimization procedure has been presented by Johnsson et al. *(15)*.

1.4. Application of Biospecific Interaction Analysis

The examples given here to illustrate BIA are collected from a series of experiments concerning structure–function analysis of recombinant human granulocyte-macrophage colony-stimulating factor (rhGM-CSF) recently conducted at our laboratory *(17)*. The rhGM-CSF molecule is a 14.7 kDa (127-amino acids) protein. The tertiary structure of the protein consists of an open bundle of four α-helices *(18,19)*. The important regions for the function of the molecule have been identified with ELISA (enzyme-linked immunosorbent assay) to appear in the first and third α-helix *(20)*, which are distant in the primary sequence but close to each other in the folded structure. Monoclonal antibodies (MAbs) directed against the entire molecule, as well as against synthetic peptides corresponding to specific sequences of the first and third α-helix, have been produced to investigate the important epitopes of the molecule.

1.4.1. Epitope Mapping of rhGM-CSF

Epitope mapping is a technique to determine the relative binding pattern of antibodies to the corresponding antigen, i.e., to investigate whether antibodies interfere in binding to the active epitopes of the molecule *(21,22)*. In total, 11

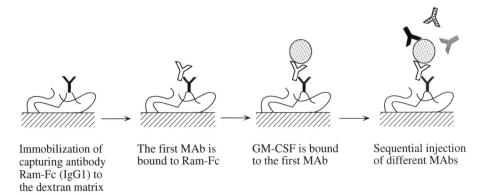

| Immobilization of capturing antibody Ram-Fc (IgG1) to the dextran matrix | The first MAb is bound to Ram-Fc | GM-CSF is bound to the first MAb | Sequential injection of different MAbs |

Fig. 5. Procedure used for determining the relative binding pattern of antiprotein MAbs to rhGM-CSF.

MAbs (antipeptide and antiprotein) were tested in this study. Two different strategies were considered for the immobilization of the rhGM-CSF molecule: (1) direct immobilization of rhGM-CSF in the matrix using amine coupling; and (2) immobilization of a capturing antibody that is used to bind the first MAb to be tested for. The rhGM-CSF molecule is then bound to this captured MAb. The first method turned out to give very irreproducible readings. This is most likely as a result of blocking of one or several epitopes during the immobilization procedure. A few of the NH_2 groups used in the amine-coupling sequence are probably located near the epitope(s), which then become sterically deactivated and thereby unable to bind to the subsequently injected MAbs. The second method appears more reliable since the binding of the protein to the surface occurs via a true recognition process. The overall mapping sequence is illustrated in **Fig. 5.** In the example below we used the amine-coupling procedure (**Subheading 3.1.**) to immobilize a capturing rabbit antimouse IgG1-Fc antibody (Ram.Fc). Ram.Fc 60 µg/mL in 10 m*M* Na-acetate buffer at pH 4.5, 35 µL for 7 min was injected over the sensing matrix, preactivated according to **steps 1–4, Subheading 3.1.** The protein rhGM-CSF is then injected and allowed to bind to the first MAb. Finally, the different MAbs (100 µg/mL in HBS, 35 µL for 7 min) are sequentially injected 100 µg/mL in HBS (35 µL for 7 min) over the sensing surface and the relative binding pattern is determined. Each injection pulse is separated by a rinsing pulse in HBS for 2 min. Up to four MAbs were analyzed sequentially.

1.4.2. Analysis

Figure 6 shows a section of a sensorgram in which the relative binding pattern of six antiprotein MAbs, emanating from two different clones, G7 and G8, were studied. The checkpoint 1 refers to the baseline obtained after immobili-

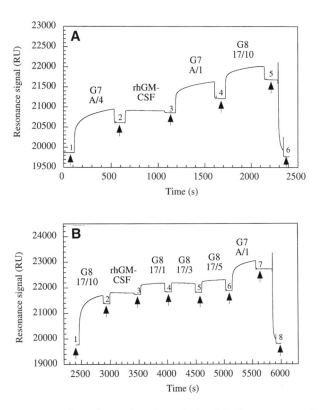

Fig. 6. Sensorgram used to determine the relative binding pattern of antiprotein MAbs to rhGM-CSF. Two short pulses of HCl, pH 2.0, are injected after the complete sequences in **(A)** and **(B)** to reach the initial baseline.

zation of Ram.Fc. The MAbs in the first series all bound with a clear response, indicating that they recognized different epitopes on the rhGM-CSF molecule (**Fig. 6A**). A change of >100 RU was taken as a positive response for the recognition process. The entire antigen–antibody complex was then dissociated using two short regeneration pulses of HCl, pH 2.0 (the onset of each pulse is indicated by the spikes between checkpoints 5 and 6 in **Fig. 6A**), and a new sequence was injected. In the second sequence the order was partly reversed and the last MAb used in the first sequence (G8.17/10) was now captured by Ram.Fc (**Fig. 6B**). If MAbs from different subclones of the G8.17 clone are sequentially injected after binding of rhGM-CSF to G8.17/10, the overall response is very low, <100 RU, except for G8.17/1 ($\approx$100 RU), suggesting that these MAbs recognize the same or partly overlapping epitopes. The sequence was terminated by injecting a MAb from the G7 clone, and again a strong response was observed, confirming our earlier observation that the G8 and G7

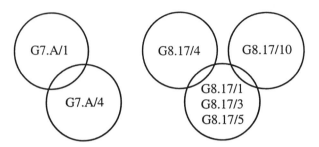

Fig. 7. Epitope map of the antiprotein MAb binding pattern to rhGM-CSF.

MAbs recognize unique epitopes on the molecule. Repeating the procedure by altering the injection sequence yields the complete epitope map of the antiprotein MAbs, as shown in **Fig. 7.** The results are presented as a map of areas representing the epitopes on the surface of the rhGM-CSF molecule. At least two distinct regions can be identified on the molecule. The first one involves the two G7 MAbs that appear to recognize different but partly overlapping epitopes. The second region displays a more complicated pattern. A common epitope is recognized by all MAbs. However, two additional epitopes appear to exist, which mutually do not interfere in binding with the corresponding MAbs, G8.17/4 (sensorgram not shown) and G8.17/10. The map in **Fig. 7** involves only the epitopes recognized by the antiprotein MAbs. A more complete analysis including the antipeptide MAbs, as well as the extracellular Rα-chain receptor, can be found in a recent study by Laricchia-Robbio et al. *(17)*.

1.5. Small Molecule Interaction and Competition Assays

It is of interest in many situations to investigate how a small molecule, e.g., a drug, interacts with a complex protein or receptor. The obvious approach is to immobilize the protein in the matrix, inject the small molecules (drug, peptide), and study the interaction. This is, however, a less favorable situation from a detection point of view. A small molecule, e.g., with a molecular weight <300, will typically give rise to a response <100 RU, a very small value (dynamic range), which in practical situations causes a large scatter in the readings. A more advantageous approach is to immobilize the small molecule in the matrix. However, not all small molecules have an amine group available for the amine-coupling procedure (**Subheading 3.1.**) An alternative immobilization protocol therefore has been developed that is based on thiol-disulfide exchange coupling. Two different methods are available: (1) ligand–thiol coupling; and (2) surface–thiol coupling. We will illustrate the first method below in which the SH group has been introduced via a cysteine at the C-terminal position of a peptide we call 14–24, which is of interest in the elucidation of the properties of rhGM-CSF.

Native Peptide

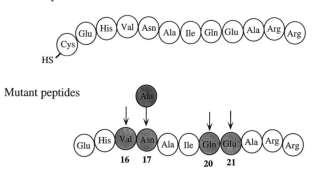

Mutant peptides

Fig. 8. Primary structures of the native and mutant synthetic peptides (14–24). In the native peptide a cysteine has been attached to the C-terminus of the sequence to ensure well-defined immobilization in the dextran matrix via thiol-disulfide exchange coupling. Mutations are introduced by alanine substitution at positions 16, 17, 20, and 21. Note that the mutant peptides do not have the cysteine at the C-terminus.

1.5.1. Competition Test Using Mutant Peptides

The first α-helix of rhGM-CSF contains an important epitope for the normal function of the molecule, which has motivated more detailed analyses of its structure–function relationship. A synthetic peptide corresponding to the amino acid sequence 14–24 was produced with cysteine attached to its low index-terminus *(20)*, **Fig. 8**, and immobilized in the matrix as can be seen in the sensorgram **Fig. 9**. Four different mutants of the peptide (14–24) were also synthesized by selective substitution of alanine at positions 16, 17, 20, and 21 *(20,24)* (**Fig. 8**) to introduce a local perturbation of the α-helix structure. These four mutant peptides were then used in a competition test with the antipeptide (14–24) MAb to highlight the critical amino acid(s) along the α-helix. The competition test is performed in the following way. Each mutant peptide was preincubated in an HBS solution of the antipeptide (14–24) MAb (clone G3.7/7) for 2 h. The solutions were then injected sequentially (35 μL for 7 min) over the sensing surface containing the immobilized native (14–24) peptide, and the presence of free antibodies was monitored with the biosensor. Each injection pulse was separated by a rinsing pulse in HBS for 2 min followed by a regeneration pulse, pH 2.0, for 1 min.

1.5.2. Analysis

The sensorgram of the mutant binding test is shown in **Fig. 10**. The solutions containing the mutant peptides with alanine in positions 20 and 21 do not cause any change in the sensorgram, indicating that there is no free, unreacted G3.7/7 MAbs left in the incubation solution. Thus, the peptides with alanine in

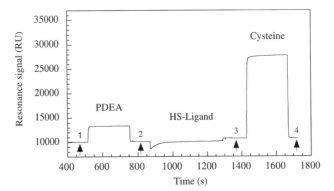

Fig. 9. Sensorgram showing the activation–immobilization–deactivation sequence used to immobilize the peptide (14–24). Checkpoint 1 refers to the baseline obtained after EDC/NHS activation. Note the small response ≈700 RU for the immobilization of the peptide (difference between checkpoints 3 and 2) as compared with that obtained for the protein in **Fig. 4.** The rapid negative step in the sensorgram after injection of the ligand is entirely the result of a difference in refractive index between the HBS and the Na-acetate buffer, $n_{\text{Na-acetate}} < n_{\text{HBS}}$.

position 20 and 21 bind strongly to the MAbs in solution, and completely inhibit further binding to the native peptide in the matrix. The sensorgrams of the MAb solutions preincubated with mutants having alanine in positions 16 and 17 reveal a completely different behavior. In particular, the sensorgram of mutant 17 displays a strong response with the biosensor, almost as strong as with the reference solution of the pure MAb, suggesting that mutation at position 17 changes the structure of the α-helix in such a way that it loses the capacity to bind the free MAb in solution. A mismatch between the structure of the binding pocket of the MAb and the peptide is obviously introduced on substitution with alanine at position 17.

The conclusion from this competition test is that the important sequence for the binding is localized to the first part (low index) of the α-helix since the structural integrity at position 17, and to some extent position 16, is of critical importance for the accessibility and presentation of the epitope. This result also agrees very well with the findings in a recent investigation using ELISA and Western blotting *(20)*.

2. Materials
2.1. Amine Coupling

1. Sensor chip CM5 (Pharmacia Biosensor AB).
2. HBS buffer: 10 nM 4-(2-hydroxyethyl)piperazine-1-ethanesulfonic acid (HEPES), 0.15 M sodium chloride, 3.4 mM EDTA, 0.005% (v/v) surfactant P-20, adjusted to pH 7.4 using NaOH.
3. 0.1 M NHS (*N*-hydroxysuccimide) adjusted to pH 8.5 with NaOH (200 μL).

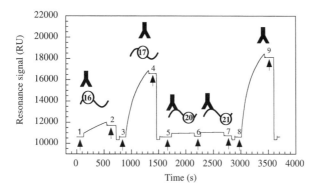

Fig. 10. Sensorgram showing the presence of free antipeptide (14–24) MAbs in solution after preincubation with mutant peptides 16, 17, 20, and 21 for 2 h in HBS buffer.

4. 0.4 *M* EDC (*N*-ethyl-*N*'-(dimethylaminopropyl)carbodiimide) adjusted to pH 8.5 with NaOH (200 μL).
5. 1 *M* Ethanolamine adjusted to pH 8.5 with NaOH (200 μL).
6. 50 μg/mL monoclonal mouse antihuman myoglobin in 10 m*M* Na-acetate buffer at pH 4.8 (100 μL).

2.2. Thiol–Disulfide Exchange Coupling

The first step of the total sequence *(23)* is identical to the activation sequence used in the amine-coupling procedure (**Subheading 3.1.**), i.e., formation of a reactive NHS ester. The subsequent steps are outlined in **Fig. 11.**

1. The sensor chip, buffer, and activation reagents were the same as for the amino-coupling kit (**Subheading 4.2.1.**).
2. 80 m*M* disulfide 2(2-pyridinyldithio)ethaneamine (PDEA), pH 8.5 (70 μL).
3. 50 m*M* cysteine in NaCl, pH 4.3 (70 μL).
4. 100 μg/mL of cysteine-modified peptide (14–24) prepared in pH 4.0, Na-acetate buffer; this pH was found to yield the highest immobilization level in a preconcentration test *(15)*.

3. Methods

3.1. The Activation–Immobilization–Deactivation Sequence

1. Degas protein, reagent, and buffer solutions. Removal of air is essential since air bubbles can cause erroneous reading of the SPR response (*see* **Subheading 4.4.3.**).
2. Equilibrate the sensor chip by letting a continuous flow (5 μL/min) of the HBS buffer pass the sample for a few minutes until a stable baseline is obtained.
3. Place the EDC, NHS, ethanolamine, and mouse antihuman myoglobin vials plus an extra mixing vial in the thermostatted (25°C) autosampler.
4. Mix 70 μL each of EDC and NHS in the extra vial, and inject the final solution (0.2 *M* in EDC and 0.05 *M* in NHS) over the sensing surface under continuous flow of 5 μL/min for 3.5 min. The activation involves the formation of an inter-

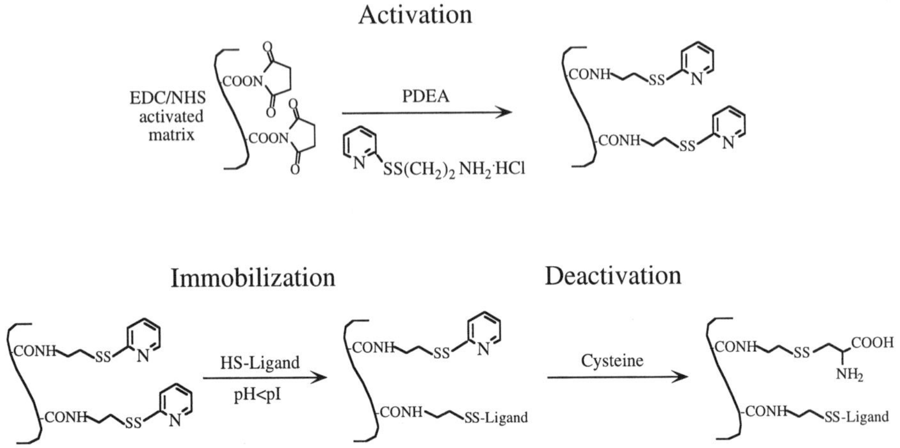

Activation

Immobilization   Deactivation

Fig. 11. The activation–immobilization–deactivation scheme for thiol-disulfide exchange coupling. The starting sequence is identical to the one used in the amine coupling kit, **Fig. 4.** The reactive disulfide group in PDEA reacts with a native or a synthetically introduced thiol group on the ligand. This method has proven particularly useful for the immobilization of small ligands, e.g., short peptides having a cysteine residue. It is also useful for immobilization of acidic proteins *(23).*

    mediate COO—EDC—urea complex, which will be substituted by a highly reactive NHS ester. The activation is then followed by rinsing in pure HBS until a stable baseline is obtained, typically for 1–2 min.

5. Inject the 35 μL of the mouse antihuman myoglobin solution over the sensing surface under continuous flow of 5 μL/min for 3.5 min. The experimental conditions employed for the immobilization procedure have been optimized according to the guidelines given by Johnsson et al. *(15).* The sequence is finally terminated by rinsing in pure HBS buffer, typically 1–2 min.

6. Deactivate the remaining NHS ester groups in the matrix by injecting a pulse of ethanolamine under continuous flow of 5 μL/min for 3.5 min. Rinse with HBS and wait for a stable baseline to appear in the sensorgram. The sensing surface is now ready for biospecific-interaction analysis with myoglobin as the antigen.

    One important advantage with the above biosensing system is that the overall activation–immobilization–deactivation process can be followed by the sensor in real-time. **Figure 12** shows the sensorgram for this automated sequence, which typically takes 30 min. The SPR-signal or resonance signal is normally given in RU (1000 RU = 1×10^{-3} refractive index units $(\Delta n_a) \approx 0.1°(\Delta \Theta sp) \approx 1$ ng/mm^2 of a protein *[11]*).

3.2. The Activation–Immobilization–Deactivation Sequence

1. Follow **steps 1–3** in the amine coupling sequence, **Subheading 3.1.**
2. Mix carefully 20 μL each EDC and NHS in an extra vial, and inject 10 μL of the final solution (0.2 *M* in EDC and 0.05 *M* in NHS) over the sensing surface under

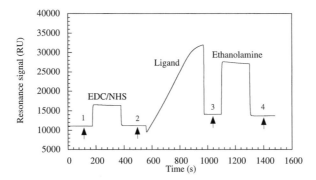

Fig. 12. Sensorgram showing the activation–immobilization–deactivation sequence used in the amine coupling kit for immobilization of mouse antihuman myoglobin.

continuous flow of 5 μL/min. The activation is terminated by rinsing in pure HBS until a stable baseline is obtained, typically for 1–2 min.

3. Inject 20 μL PDEA over the sensing surface (5 μL/min) to introduce the reactive disulfide. This second activation is followed by rinsing in pure HBS until a stable baseline is obtained, typically for 1–2 min.

4. Inject 35 μL of the peptide (14–24) under continuous flow (5 μL/min) in Na-acetate buffer adjusted to pH 4.0. Rinse with pure HBS until a stable baseline is obtained, typically for 1–2 min.

5. Deactivate excess disulfides by injecting 20 μL cysteine, pH 4.3, at a flow rate of 5 μL/min. Final rinsing in HBS is then performed until a stable baseline is obtained, typically for 1–2 min. Then the surface is ready for biospecific interaction analysis with the peptide (14–24) as the ligand.

The sensorgram showing the immobilization sequence of peptide (14–24) is illustrated in **Fig. 9.**

4. Notes

A few practical features of SPR detection and BIA are worth mentioning before going into a detailed description of the various application examples. Four topics will be briefly discussed: bulk effects in SPR detection, calibration, regeneration, and factors influencing the performance of the biosensor.

4.1. Bulk Effects in SPR Detection

The SPR response is proportional to changes in the refractive index that occur in an interaction volume outside the metal, which is defined by the penetration of the evanescent field into the ambient medium. Two effects can contribute to the change in refractive index and thus to the SPR signal: (1) replacement of buffer with biomolecules because of biospecific interaction; (2) differences in the "bulk" refractive index between the buffer and the solutions, e.g., EDC/NHS, ligand, analyte, and deactivation. How can one separate these two contributions to the SPR signal? In **Fig. 12** we observed that the SPR signal increased rapidly on injection of the activation solution. Most of this signal disappears, however, during the rinsing pulse. These fast changes in the SPR signal are characteristic for "bulk" effects and reflect only the difference in refractive index between the HBS buffer and the EDC/NHS solution. The sign of this change can be negative as well as positive depending on whether $n_{buffer} > n_{solution}$, or vice versa. A rapid change (decrease) of the SPR signal is also observed during the immobilization cycle. This rapid decrease is followed by another process occurring on a completely different timescale, more representative of electrostatic attraction of the protein to the surface, followed by interaction (binding). On rinsing with buffer the signal does not return to its initial value, indicating that a permanent change in the refractive index of the matrix has occurred, and the true SPR signal is in this particular case equivalent to the difference in readings of the baseline at points 2 and 3, respectively, about 2900 RU. Bulk effects are generally not a serious problem for the identification of a certain biomolecular interaction. However, the bulk effects must be compensated for when kinetic information and accurate concentration determination are asked for. The method normally used is schematically described in **Fig. 13**, in which the first experiment involves injection of the analyte solution over a sensing surface with the ligand present. A second sensorgram is obtained under identical conditions but without the ligand present. This sensorgram, **Fig. 13B**, is then subtracted from the sensorgram of the activated matrix, **Fig. 13A**, yielding the true sensorgram of the biospecific interaction (**Fig. 13C**).

4.2. Regeneration

Many of the interactions encountered in BIA are of noncovalent character. An efficient *in situ* regeneration procedure is therefore required so that the sensing surface with the immobilized ligand can be used repeatedly. The regeneration solution should rapidly dissociate the interaction pair and elute the analyte from the interaction matrix. At the same time the ligand should survive and not lose its activity so that it can be used again, typically for 100 analyses. The most convenient way to regenerate the dextran matrix involves exposure to a short pulse of HCl at pH <2.5. Regeneration with NaOH at pH >10.0 can also be used when acidic regeneration is unsatisfactory. However, it should be used with great care since denaturation of many proteins often occurs at high pH.

4.3. Calibration

For analyte concentration determination, the SPR signal must be calibrated. The relation between analyte concentration and the SPR signal is dependent on sample contact time, mass transport (viscosity, flow-cell dimensions, flow

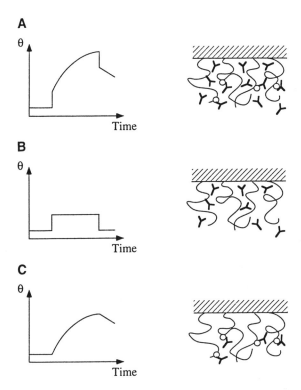

Fig. 13. Schematic outline of the methodology used to compensate for "bulk" effects in an SPR experiment. The sensorgram obtained without ligand present **(B)** is subtracted from that obtained with the ligand present **(A)** to yield the true sensorgram of the interaction process **(C)**.

rate), immobilized ligand concentration, steric hindrance (a function of bound analyte), temperature, and refractive index of the buffer. There is no need to know each parameter; the calibration can be done with standard solutions at a few discrete concentration points. A curve-fitting process, e.g., cubic spline, will generate a calibration or standard curve.

For simple kinetic analysis there is no need to calibrate the SPR signal as long as it is proportional to the bound analyte surface concentration. Because of small response levels, the above condition is often fulfilled. Complex kinetics, e.g., two-state binding, may require calibration.

4.4. Factors Influencing the Performance of the Biosensor

4.4.1. Level of Immobilization

The desired level of immobilization depends on the type of experiment performed. A rule of thumb is that for concentration measurements the immobilization level should be as high as possible to ensure efficient binding of the analyte. For kinetic measurements the immobilization level should be as low as possible to avoid diffusion-limited processes. Note that the immobilization level in this case means the concentration of the ligand (mol/mm^2), not necessarily the response ($\Delta\Theta_{sp}$), which is dependent on the molecular weight of the ligand.

4.4.2. Temperature

A temperature change during a measurement will cause a drift because the buffer will change refractive index. The change is about 1×10^{-4} refractive index U/C° corresponding to approx 100 RU/C°.

4.4.3. Air Bubbles

It is very important to use degassed solutions, especially for the running buffer. Even if the buffer is degassed, the ambient air will continuously be dissolved in the buffer. It is recommended to degas or change buffer every 12 h. Air dissolved in the buffer may cause air bubbles in the flow system, which can lead to large errors in the SPR signal. If the analysis temperature is higher than the ambient temperature, the buffer is more prone to create air bubbles.

4.4.4. Clogging

Biomolecule and salt residues can clog the flow system if not washed with deionized water.

4.4.5. Flow Rate

For concentration measurements a low flow rate is desired for small sample volumes. Kinetic measurements may require high flow rates to avoid diffusion limitations. Flow rates, that are too low, e.g., <2 µL/min, may cause an unstable flow. Flow rates that are too high may cause temperature drifts and high shear rates at the sensor surface. Commonly used flow rates are between 5 and 20 µL/min.

4.4.6. Sensitivity and Dynamic Range

The detection limit of the biosensing system is approx 1 RU, which corresponds to 1 pg/mm^2 of biomolecules. The dynamic range in refractive index is 1.33–1.36, which corresponds approx to 30 ng/mm^2.

5. Summary

Affinity biosensing with surface plasmon resonance detection has developed into a versatile tool for real-time biospecific-interaction analysis. In this chapter we have demonstrated how the technology can be used for epitope mapping and structure–function analysis of synthetic peptide–MAb interactions. Concentration determination and kinetic analysis are other important application areas of BIA, and publications covering these topics can be found elsewhere *(25–28)*.

Acknowledgments

We thank L. Laricchia-Robbio and R. Revoltella, Institute of Mutagenesis and Cell Differentiation, CNR, Pisa, Italy, for a fruitful collaboration on the structure–function analysis of rhGM-CSF. This work was supported by the Swedish Research Council for Engineering Sciences, Pharmacia Biosensor AB and the European Science Foundation (ESF) through the program Artificial Biosensing Interfaces (ABI). We also thank Ingemar Lundström for critical reading of the manuscript.

References

1. Jönsson, U., Fägerstam, L., Ivarsson, B., Johnsson, B., Karlsson, R., Lundh, K., Löfås, S., Persson, B., Roos, H., Rönnberg, I., Sjölander, S., Stenberg, E., Ståhlberg, R., Urbaniczky, C., Östlin, H., and Malmqvist, M. (1991) Real-time biospecific interaction analysis using surface plasmon resonance and a sensor chip technology. *Bio/Techiques* **11**, 620–627.
2. Fägerstam. L. (1991) A non label technology for real-time biospecific interaction analysis, in *Techniques in Protein Chemistry*, vol. 2 (Villafranca, J. J., ed.), Academic, New York, pp. 65–71.
3. Jönsson, U. and Malmqvist, M. (1992) Real time biospecific analysis. *Adv. Biosensors* **2**, 291–336.
4. Liedberg, B., Nylander, C., and Lundström, I. (1983) Surface plasmon resonance for gas detection and biosensing. *Sensors Actuators* **4**, 299–304.
5. Raether, H. (1988) *Surface Plasmons on Smooth and Rough Surfaces and on Gratings*. Springer-Verlag, Berlin.
6. Boardman, A. D. (1982) *Electromagnetic Surfaces Modes*, Wiley, Chichester, UK.
7. Swalen, J. D., Gordon, J. G., Philpott, M. R., Brillante, A., Pockrand, I., and Santo, R. (1980) Plasmon surface polariton dispersion by direction optical observation. *Am. J. Phys.* **48**, 669–672.
8. Kretschmann, E. (1971) Die Bestimmung optischen Konstanten von Metallen durch Anregung von Oberfläschenplasmaschwingen. *Z. Phys.* **241**, 313–323.
9. de Feijter, J. A., Benjamins, J., and Veer, F. A. (1978) Ellipsometry as a tool to study the adsorption of synthetic and biopolymers at the air-water interface. *Biopolymers* **17**, 1759–1772.
10. Mayo, C. S. and Hallock, R. B. (1989) Immunoassay based on surface plasmon resonance, *J. Immunol. Meth.* **120**, 105–114.
11. Stenberg, E., Persson, B., Roos, H., and Urbaniczky, C. (1991) Quantitative determination of surface concentration of protein with surface plasmon resonance by using radio labelled proteins, *J. Colloid Interface Sci.* **143**, 513–526.
12. Löfås, S., Malmqvist, M., Rönnberg, I., Stenberg, E., and Liedberg, B., and Lundström, I. (1991) Bioanalysis with surface plasmon resonance. *Sensors Actuators B* **5**, 79–84.

13. Liedberg, B., Stenberg, E., and Lundström, I. (1993) Principles of biosensing with an extended coupling matrix and surface plasmon resonance. *Sensors Actuators B* **11,** 63–72.

14. Löfås, S. and Johnsson, B. (1990) A novel hydrogel matrix on gold surfaces in surface plasmon resonance sensors for fast and efficient covalent immobilization of ligands, *J. Chem. Soc. Chem Commun.* 1526–1528.

15. Johnsson, B., Löfås, S., and Lindquist, G. (1991) Immobilization of proteins to carboxy methyl dextran-modified gold surfaces for biospecific analysis in surface plasmon resonance sensors. *Anal. Biochem.* **198,** 268–277.

16. Sjölander, S. and Urbaniczky, C. (1991) Integrated fluid handling system for biomolecular interaction analysis. *Anal. Chem.* **63,** 2338–2345.

17. Laricchia-Robbio, L., Liedberg, B., Platou-Vikinge, T., Rovero, P., Beffy, P., and Revoltella, R. P. (1996) Mapping of monoclonal antibody- and receptor-binding domains on human granulocyte-macrophage colony stimulating factor (rhGM-CSF) using a surface plasmon resonance-based biosensor. *Hybridoma* **15,** 343–350.

18. Diederichs, K., Boone, T., and Karplus, P. A. (1991) Novel fold and putative receptor binding site of granulocyte-macrophage colony-stimulating factor. *Science* **265,** 1779–1782.

19. Walter, M. R., Cook, W. J., Ealick, S. E., Nagabhushan, T. L., Trotta, P. P., and Bugg, C. E. (1992) Three-dimensional structure of recombinant human granulo-zyte-macrophage colony-stimulating factor. *J. Mol. Biol.* **224,** 1075–1085.

20. Beffy, P., Rovero, P., Di Bartolo, V., Laricchia-Robbio, L., Dané, A., Pegoraro, S., Bertolero, F., and Revoltella, R. (1994) An immunodominant epitope in a functional domain near the N-terminus of human granulocyte-macrophage colony-stimulating factor identified by cross-reaction of synthetic peptides with neutralizing anti-protein and anti-peptide antibodies. *Hybridoma* **13,** 457–468.

21. Fägerstam, L. G., Frostell, Å., Karlsson, R., Kullman, M., Larsson, A., Malmqvist, M., and Butt, H. (1990) Detection of antigen-antibody interactions by surface plasmon resonance. Application to epitope mapping. *J. Mol. Recognition* **3,** 208–214.

22. Nice, E., Layton, J., Fabri, L., Hellman, U., Engström, Å., Persson, B., and Burgess, A.W. (1993) Mapping of the antibody– and receptor binding domains of granulocyte colony-stimulating factor using an optical biosensor. Comparison with enzyme-linked immunosorbent assay competition studies. *J. Chromatogr.* **646,** 159–168.

23. Application note 601, Ligand immobilization for real-time BIA using thiol-disulphide exchange, Pharmacia Biosensor, Uppsala, Sweden.

24. Beffy, P., Di Bartolo, V., Laricchia-Robbio, L., Pegoraro, S., Chiello, E., Rovero, P., Caracciolo, L., and Revoltella, R. (1994) Small synthetic peptides of human GM-CSF require different conditions for immobilization, epitope density and presentation in ELISA. *Fund. Clin. Immunol.* **2,** 53–61.

25. Fägerstam, L. G., Frostell-Karlsson, Å., Karlsson, R., Persson, B., and Rönnberg, I. (1992) Biospecific interaction analysis using surface plasmon resonance detection applied to kinetic, binding site and concentration analysis, *J. Chromatog.* **597,** 397–410.

26. Karlsson, R., Fägerstam, L., Nilshans, H., and Persson, B. (1993) Analysis of active antibody concentration. Separation of affinity and concentration parameters. *J. Immunol. Meth.* **166,** 75–84.

27. Karlsson, R., Michaelsson, A., and Mattsson, L. (1991) Kinetic analysis of monoclonal antibody-antigen interactions with a new biosensor based analytical system. *J. Immunol. Meth.* **145,** 229–240.
28. Karlsson, R. (1994) Real-time competitive kinetic analysis of interactions between low-molecular-weight ligands in solution and surface-immobilized receptors. *Anal. Biochem.* **221,** 142–151.

4

Immunosensors Based on Piezoelectric Crystal Device

Marco Mascini, Maria Minunni, George G. Guilbault, and Robert Carter

1. Introduction

New recent developments in engineering have improved transducer piezoelectric technology and have led to a new generation of sensor devices, the piezoelectric microbalances, based on planar microfabrication technique. These devices show a very high sensitivity (up to femtomole) for detecting molecules that are linked to the surface of the device and change its resonant frequency.

A sensor surface can be coated selectively for interaction with a specific chemical or class of chemical or a biomolecule to be detected. Applications in medicine, environmental monitoring, food analysis, and process control are possible. Most of the applications employ bulk acoustic waves (BAW) devices, using a selective coating for the analyte of interest.

The elucidation of a linear relationship between the change in the oscillating frequency of a piezoelectric crystal and the mass variation on its metallic surface has produced diverse studies and applications of piezoelectric sensors in many areas of analytical chemistry. The discovery of piezoelectricity itself by Jacques and Pierre Curie dates back to 1880; they discovered that a mechanical stress applied to surfaces of various crystals, including quartz, Rochelle salt, and tourmaline, afforded a corresponding electrical potential across the crystal whose magnitude was proportional to the applied stress. This behavior is referred to as the piezoelectric effect, which is derived from the Greek word *piezein*, meaning to press *(1)*. The piezoelectric effect is a result of the fact that certain crystals contain positively and negatively charged ions that separate when the crystals are subjected to stress. The mechanical shear stress results in a separation of charge centers called polarization. Any given crystal has a

From: *Methods in Biotechnology, Vol. 7: Affinity Biosensors: Techniques and Protocols*
Edited by: K. R. Rogers and A. Mulchandani © Humana Press Inc., Totowa, NJ

natural vibration frequency, called its resonant (or fundamental) frequency, which depends on its chemical nature, its size and shape, and its mass *(2)*.

When the vibrating crystal is piezoelectric, this cycle of oscillating deformity produces an oscillating electrical field; the frequency of the electrical oscillation is identical to the vibration frequency of the crystal. At the same time, placing a piezoelectric crystal in an oscillating electrical field causes it to vibrate at the frequency of the oscillating field. This transfer of energy from the electrical field to the crystal is very inefficient except when the frequency of the oscillating electrical field is the same as the resonant frequency of the crystal. This is exploited by the incorporation of the quartz crystal into oscillator circuits, with the frequency of the entire circuit becoming the resonant frequency of the quartz crystal. From the same principle, the resonant frequency of a given crystal, for instance a crystal used as a biosensor, can be determined from the frequency of an electrical oscillator circuit in which the crystal is a component.

Electrodes are used to apply the electric field across the crystal. The resulting stress can result in different modes of oscillation depending not only on the parameters mentioned above but also on the configuration of the electrode used to induce the field. As with all mechanical structures, a piezoelectric crystal resonator (which is a precisely cut slab from a natural or synthetic crystal of quartz) can have many modes of resonance, or standing-wave patterns at the resonant frequencies. As a rectangular solid bar, for example, it may exhibit three different types of vibrations: longitudinal (extentional), lateral (flexural or shear), and torsional (twist), in each of the three axes. In addition to the fundamental modes, the system can also vibrate at the overtones of each fundamental mode. Several modes may also couple to form very complicated resonance modes. In general, one would like to have the quartz resonator oscillating at only one principal mode. The selection of one particular mode and the suppression of all unwanted modes require that the crystal slab be cut at a specific crystallographic orientation and have the proper shape. The configuration of the electrodes on the quartz crystal resonator, the supporting structure, and the oscillating circuit may also significantly affect the mode of resonance. For example, AT-cut quartz crystal is obtained by cutting the wafer of quartz at approx 35° from the z-axis. Application of an alternating electrical field across the thickness of an AT-cut quartz crystal by two excitation electrodes on opposite sides of the crystal results in a shear vibration (**Fig. 1**) in the x-axis direction parallel to the electrical field and propagation of a transverse shear wave through the crystal in the thickness direction.

The principal piezoelectric materials used commercially are crystalline quartz and Rochelle salts. Quartz has the important quality of being a completely oxidized compound (silicon dioxide). It is insoluble in water and resists

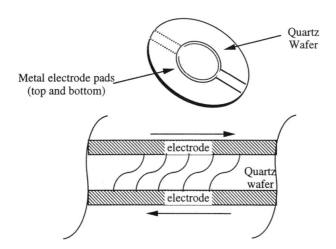

Fig. 1. A typical piezoelectric crystal orientation of electrodes and shear deformation during oscillation at the fundamental frequency.

temperatures of up to 579°C with no loss of piezoelectric properties. Industrially grown rather than natural quartz crystals are used almost exclusively for electrogravimetric sensors because of their higher purity.

The piezoelectric immunosensor consists of a piezoelectric crystal with an antigen or antibody immobilized on its surface. The biospecific reaction between the two interacting molecules, one immobilized on the surface and the other free in solution, can be followed in real-time without the use of any label. The piezoelectric crystal is a mass device; this means that any surface mass change reflects on its resonant frequency. The increase of mass at the sensor surface results in a decrease of the frequency as shown by Sauerbrey, who first introduced the quartz crystal microbalance principles *(3)*. This relationship of frequency to mass has been outlined by the following equation:

$$\Delta f = -2.3 \times 10^{-6} \, F^2 \, \Delta m / A \qquad (1)$$

where Δf is the change in the fundamental frequency of the coated crystal, F is the resonant frequency of the crystal, A is the area of the electrode coated, and Δm is the change in mass resulting from material binding to the crystal. The oscillation frequency is influenced by some factors that remain constant, such as the thickness, the density, and the shear modulus of the quartz, and by the physical parameters of the adjacent media (density or viscosity of air or liquid). Many researchers have investigated this problem and proposed experimental and theoretical equations describing the oscillation frequency in solution *(4–10)*.

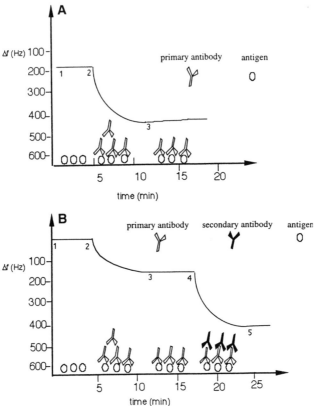

Fig. 2. Direct assay. **(A)** Single step: The buffer flows over the surface with immobilized antibodies, the relative antigen is added and binds the antibody, and the surface is washed with buffer to remove the excess antigen. **(B)** Multistep: The buffer flows over the surface with immobilized antibodies, the relative antigen is added and binds the antibody, the surface is washed with buffer to remove the excess antigen, a secondary antibody against antigen is added to enhance the first binding, then the surface is washed with buffer to remove the excess secondary antibody.

The piezoelectric immunosensor can use single- or multistep binding to the crystal surface and direct or indirect measurement of analyte.

1. Single-step method measures the binding of one component to the modified crystal surface. Multistep methods rely on sequential binding of two or more components (**Figs. 2** and **3**).
2. Direct measurement relies on interaction of the analyte itself with the modified crystal surface (i.e., IgG determination using immobilized anti-IgG antibodies). This response increases with increasing amount of analyte.
3. Indirect measurement relies on the interaction of the analyte with some other components free in solution (i.e., determination of the herbicide 2,4-dichlorophenoxyacetic acid).

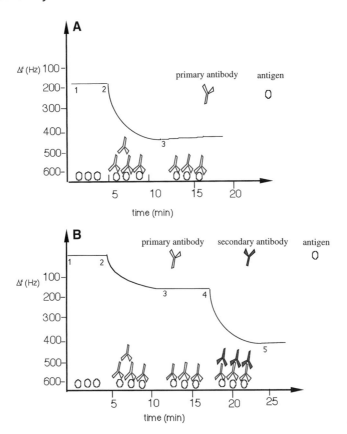

Fig. 3. Indirect assay. **(A)** Single-step: The buffer flows over the surface with immobilized antigen, the relative antibody is added and binds the antigen, and the surface is washed with buffer to remove the excess antibody. **(B)** Multistep: The buffer flows over the surface with immobilized antigen, the relative antibody is added and binds the antigen, and the surface is washed with buffer to remove the excess antibody. A secondary antibody against the primary antibody is added to enhance the first binding, then the surface is washed with buffer to remove the excess secondary antibody.

In the competitive immunoassay the antigen is immobilized on the crystal surface and the analyte present in solution competes for the binding sites of the antibody with the antigen immobilized. The observed frequency shift is inversely related to the analyte concentration (as reported for 2,4-D).

In the displacement assay, a particular competition assay, the antigen is immobilized on the crystal surface (*see* example for *Listeria monocytogenes* reported in **Subheading 3.2.3.**). The addition of the related antibody results in the formation of the immunocomplex with an increase in mass and a corresponding frequency decrease. The antigen present in the added sample competes for the antibody site previously bound to the crystal surface and causes their displacement with a resulting increase in frequency. The displacement of antibody from the surface is proportionally related to the concentration of the analyte in the sample.

Every change in the surface concentration is registered as a frequency change of the crystal. Both direct and indirect methods involve the immobilization of one interactant on the crystal surface. A wide range of crystals is commercially available but probably the most commonly used crystals are from Universal Sensors, Inc. (Metairie, LA) and Seiko (Chiba, Japan). These consist of a quartz wafer on which is mounted an electrode (silver, platinum, or gold) vaporized on the crystal's surface.

The immobilization procedure can be performed in many ways depending on whether the capturing molecule should be attached covalently or simply adsorbed on the crystal surface or even entrapped into polymers. Regardless of the coupling method, the goal is to form a local concentration of the affinity ligand across a biospecific surface. Correct orientation and retention of activity are important in this process, especially for ligands containing active sites that must interact with specific analytes after immobilization *(11)*.

In general, adsorption procedures do not yield stable affinity systems for biosensor design because the weak bonds created by noncovalent attachment usually causes severe leakage of the biomolecule off the surface and degradation of performance with use. An intermediate approach is to adsorb affinity-binding proteins, like the material protein A or protein G, that are able to bind the Fc portion of the antibodies. The binding of the antibodies to the matrix surface occurs in an oriented manner. This enhances the surface binding capacity since the Fab antibody portion (the recognition site) is available for the antigen binding *(12)*.

The same idea is behind the avidin adsorption on the sensor surface. The avidin is adsorbed on the crystal; biotinylated antibodies can then be coupled to the surface via the high-affinity avidin–biotin binding (K_A 10^{15}).

For small ligands, as well as for proteins, covalent coupling provides the best approach to immobilize molecules. The covalent immobilization technique requires activation chemistry. The surface should be activated to allow the coupling of the appropriate chemical groups on the target ligand. The chemistry most often used employs water-soluble carbodiimides, e.g., 1-ethyl-3(3-dimethylamino-propyl)carbodiimide (EDC), when an amino group and a carboxyl group are available on the two ligands, or glutaraldehyde to couple two amino groups, and even other bifunctional compounds that have two reactive portions.

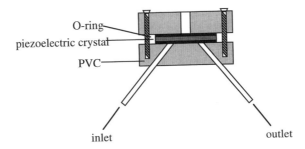

Fig. 4. Flow cell. The cell is composed of acrylic, with an upper piece and a lower piece held together with two nylon thumb screws. The crystal is centered between two O-rings in the upper and lower pieces. The cell allows the use of liquid samples with piezoelectric crystals in two methods. One side of the cell is constructed as a static system. In this case, one side of the crystal is exposed to a chamber that can hold up to 1 mL liquid. The other side of the cell is constructed as a flow system. In this case, one face of the crystal is exposed to a 40-µL chamber. This chamber can be connected to an external peristaltic pump.

2. Materials

2.1. Common to All Assays

Use pure-grade reagents and distilled water for all solutions.

1. The apparatus required is a piezoelectric detector, available commercially (Universal Sensors; Seiko), which includes a frequency meter, an oscillator, a digital output for connection to computer, and software for collecting and displaying data from the instrument (PZTools© software from Universal Sensors).
2. AT-cut piezoelectric crystals (from 5 to 10 MHz, Universal Sensors).
3. A cell for static or flow measurements. **Figure 4** shows the Universal Sensors model.
4. A peristaltic pump.
5. Crystal cleaning solution: 1.2 N NaOH and 1.2 N HCl; concentrated HCl;
6. Coupling buffer: 10 mM phosphate buffer, pH 7.4. Dissolve 445 mg Na_2HPO_4 · $2H_2O$ in 200 mL water, adjust the pH to 7.4 by addition of HCl, and make up to 250 mL in water for use.

2.2. Silanization (A)

1. 5% 3-Aminopropyltriethoxy-silane (APTES, Sigma, St. Louis, MO) in acetone. Add 250 µL APTES to 4750 µL acetone (for 5 mL final volume).
2. 2.5% Glutaraldehyde (GA) in 100 mM phosphate-buffered saline (PBS), pH 7.4 buffer. For 100 mM PBS: Dissolve 445 mg Na_2HPO_4 · $2H_2O$ in 20 mL water, adjust to pH 7.4 by addition of HCl, and make up to 25 mL in water. Take 5 mL 25% GA and dilute 1:5 with PBS (add 20 mL 100 mM PBS to 5 mL GA).

3. Coupling molecule solution: Weigh 1 mg of binding molecule (BSA) and add 1 mL PBS 10 mM, pH 7.4, to obtain 1 mg/mL concentration.
4. 100 mM glycine to saturate the activated sites that did not react with coupling molecule.
5. Regenerating agent: 10 mM NaOH, 10 mM HCl, or glycine solution, pH 2.5.

2.3. Protein A (or Protein G) Coupling

1. Protein A (or protein G solution) from Pierce (Rockford, IL): Aliquot 1 mg/mL solution in 50-μL vials in Dulbecco buffer, pH 7.4 (*see* **item 2**).
2. Buffer: Dulbecco's phosphate-buffered saline (DPBS) is composed of 137 mM NaCl, 2.7 mM KCl, 8.0 mM Na$_2$HPO$_4$, and 1.5 mM KH$_2$PO$_4$, pH 7.4.
3. Coupling molecule, i.e., 1 mg/mL antibodies in 10 mM DPBS buffer, pH 7.4, or for the competitive assay, as in our case, *Listeria* heat-killed cells, 10^9 cells/mL, (Kirkegaard and Perry, Gaithersburg, MD). Take the vial with *Listeria* and add 500 μL buffer. The final concentration is then 10^9 cells/mL. Aliquot in 50 μL and store frozen at –20°C.
4. Anti-*Listeria* antibodies: 1 mg/mL in Dulbecco buffer, pH 7.4.

2.4. Avidin-Biotin Coupling

1. Avidin solution: Aliquot 1 mg/mL solution in 50-μL vials. Dilute the 1 mg/mL solution 1:5 by adding 50 μL to 450 μL 10 mM PBS, pH 7.4.
2. Biotinylated antibodies: antihuman IgG antibodies from Sigma (0.2 mg/mL).

3. Methods
3.1. Immobilization Procedure
3.1.1. Common to All Assays: Crystal Cleaning

1. Take the crystal and dip it in a beaker containing 5 mL 1.2 N NaOH for 5 min; wash with 5 mL distilled water.
2. Dip the crystal into 5 mL 1.2 N HCl for 5 min; add 10 μL concentrated HCl on the gold electrodes, taking care to keep the electrode contacts out of the solution. After 2 min wash the crystal again and let it dry.

An alternative cleaning is:

1. Take the crystal and dip it into a beaker containing a mixture of H$_2$O, H$_2$O$_2$, and NH$_4$OH (ratio 5:1:1) at 80°C for 5 min.
2. Rinse with 10 mL Millipore-filtered water or distilled water *(13)*.

3.1.2. For Silanization

1. Dip the crystal into 5 mL APTES 5% acetone solution for 1.5 h.
2. Wash with 5 mL water and then dry at 100°C for 1 h.
3. Dip the crystal into 5 mL 2.5% GA, 100 mM PBS solution, pH 7.4, then wash it with distilled water and air dry it.

4. Add the immobilizing compound (5 μL side of the crystal), then leave for 1 h (for proteins).
5. The crystal is then dipped into 5 mL 100 m*M* glycine solution for 1 h, washed with 5 mL distilled water, and air dried.

3.1.3. Protein A (or Protein G) Coupling

1. Insert the crystal into the cell, connect the electrodes of the crystal to the detector using a shielded cable, and record the frequency with the computer.
2. Add 5 μL protein A (or protein G) solution, 1 mg/mL. Cover the upper cell with Parafilm™ to prevent evaporation and then leave at 37°C for 1 h; wash with 500 μL of the same buffer three times. Add 20 μL the coupling molecule, i.e., 1 mg/mL antibodies or antigen in the case of a competitive assay (bacterial cells, in our case, *Listeria* heat-killed cells, 10^9 cells/mL).

3.1.4. Avidin-Biotin Coupling

1. Insert the crystal into the cell, connect the electrodes of the crystal to the detector using a shielded cable, and record the frequency with the computer.
2. Add 25 μL of 10 m*M* PBS buffer, pH 7.4, to equilibrate the surface; the frequency decreases until it gets stable.
3. Aspirate the buffer with the same pipet.
4. Change the tip and add 25 μL avidin solution (200 μg/mL) in PBS. The frequency decreases because the protein avidin is adsorbed to the surface; after the frequency stabilization (wait at least 30 min) wash the surface with 25 μL PBS buffer and take the frequency value.
5. Aspirate the buffer with the same pipet.
6. Add 25 μL biotinylated antibody (0.2 mg/mL). A frequency decrease is observed; after 20 min aspirate the solution, wash the surface with 25 μL buffer, and take the frequency.
7. Wash with 100 μL buffer and add 200 μL human immunoglobulin (h-IgG) solution at different concentrations.

3.2. Assay Development

In the case of the silanization, after the immobilization insert the crystal into the cell. In the case of avidin–biotin coupling the cell is already mounted with the crystal. After the immobilization has been performed the amount of immobilized material can be calculated from the frequency decrease subtracting the original frequency before immobilization from the last frequency observed. In the case of avidin adsorption the frequency shift is about 250 Hz.

Then the interacting molecule is added. The resulting frequency shift is proportional to the sample antigen concentration.

We report here three different approaches used for h-IgG determination, pesticide detection (herbicide 2,4-D) as an example of small molecule analyte, and bacterial cell detection (*Listeria*).

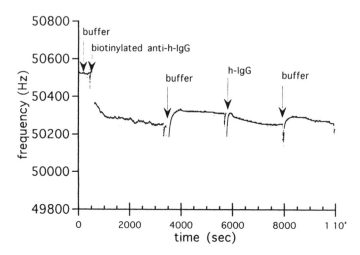

Fig. 5. Real-time monitoring of the interaction of biotinylated antihuman IgG with the crystal surface previously coated with avidin. Here a direct assay is performed: The crystals are incubated in the presence of biotinylated anti-h-IgG antibodies (0.20 mg/mL) and 200 µL of different concentrations of h-IgG (sample) for a limited amount of time (20 min for each concentration) are added after washing with 200 µL distilled water. The solution containing the sample comes in contact with the surface and the h-IgG free in the sample binds the anti-h-IgG immobilized on it. After the interaction, the buffer replaces the antigen (h-IgG) solution and the frequency value after the washing is taken. The frequency shift obtained after the antibody–antigen interaction is proportional to the concentration of free antigen (h-IgG).

3.2.1. Application 1: Human Immunoglobulin (h-IgG) Determination

The cell is used in the static configuration. Surfaces with adsorbed avidin have been exposed to a solution of h-IgG.

1. Add 200 µL of a solution of biotinylated antihuman IgG antibodies (200 µg/mL in 10 mM PBS, Sigma). Avidin anchors the anti-h-IgG via biotin.
2. Wash the surface with 200 µL 10 mM PBS.
3. Aspirate the buffer and add the h-IgG solution (Sigma, St. Louis, MO) *(14)*. A direct assay is performed (*see* **Fig. 2A**). Different concentrations from 10 to 200 µg/mL of h-IgG solutions are added to the surface. After each addition a washing step is performed with 200 µL 10 mM PBS and the frequency recorded (**Fig. 5**).
4. The surface is regenerated using 0.1 M glycine, pH 2.1, for 3 min.

In the direct assay the crystals are incubated in the presence of anti-h-IgG antibodies (20 µg/mL) as shown in **Fig. 5**. There is a correlation between the frequency shift and the amount of analyte in the sample. The analysis time is 15 min for each measurement.

3.2.2. Application 2: Determination
of the Pesticide Dichlorophenoxyacetic Acid (2,4-D)

Silanized surfaces, obtained as described in **Subheading 3.1.2.**, have been extensively used for the immobilization of both small and big molecules *(15)*. An indirect competitive assay is performed (*see* **Fig. 3A**); the 2,4-D antigen is immobilized on the silanized surface via coupling with bovine serum albumin (BSA). After the GA step (**Subheading 3.1.2., step 3**) the following procedures are done:

1. Add 20 µL/side BSA 50 mg/mL in 100 mM phosphate buffer, pH 7.4) to the surface; the amino groups of the BSA are then used for the 2,4-D coupling.
2. Wash the crystal with distilled water.
3. Dip the crystal in 5 mL of previously activated 2,4-D solution and incubate overnight; wash and store at 4°C. Modification of 2,4-D for coupling is reported below.

3.2.2.1. MODIFICATION OF 2,4-D FOR COUPLING

1. Dissolve 300 mg of 2,4-D in 7 mL of dioxane (Merck, Darmstadt, Germany).
2. Add 600 µL of tributylamine (Serva, Heidelberg, Germany) to the solution and cool in ice bath at −10°C.
3. While stirring slowly add 150 µL isobutylchloroformate (Serva). Stir the solution for 30 min.
4. Add 25 mL cold dioxane, 35 mL water, and 3 mL 1.2 N NaOH (resulting in a pH of 10.0–13.0).
5. Apply this solution to the crystal. The chemistry of the reaction and the chemically modified surface are shown in **Scheme 1**.

3.2.2.2. BINDING OF THE ANTIBODY TO THE SURFACE WITH THE IMMOBILIZED ANTIGEN

Figure 6 shows the interaction of the anti-2,4-D antibody in solution with the 2,4-D immobilized on the surface.

1. Add 500 µL of a solution of 1 mg/mL anti-2,4-D monoclonal antibodies (MAbs kindly supplied by M. Franek laboratory, Dept. of Biochemistry, Masaryk University, Brno, Czech Republic) to the cell.
2. After binding, wash the surface with 500 µL 10 mM PBS to remove the excess antibody. The affinity reaction is relatively fast; 10 min. A control for unspecific binding with h-IgG is performed.
3. Add 1 mg/mL h-IgG solution in 10 mM PBS and after 10 min wash with buffer (10 mM PBS). The h-IgG solution does not give any frequency shift (data not shown).

The kinetics of the process can be followed in real-time, and two different clones (1 and 2), at a concentration of 1 mg/mL, can be compared in their ability to bind the surface as shown in **Fig. 7**.

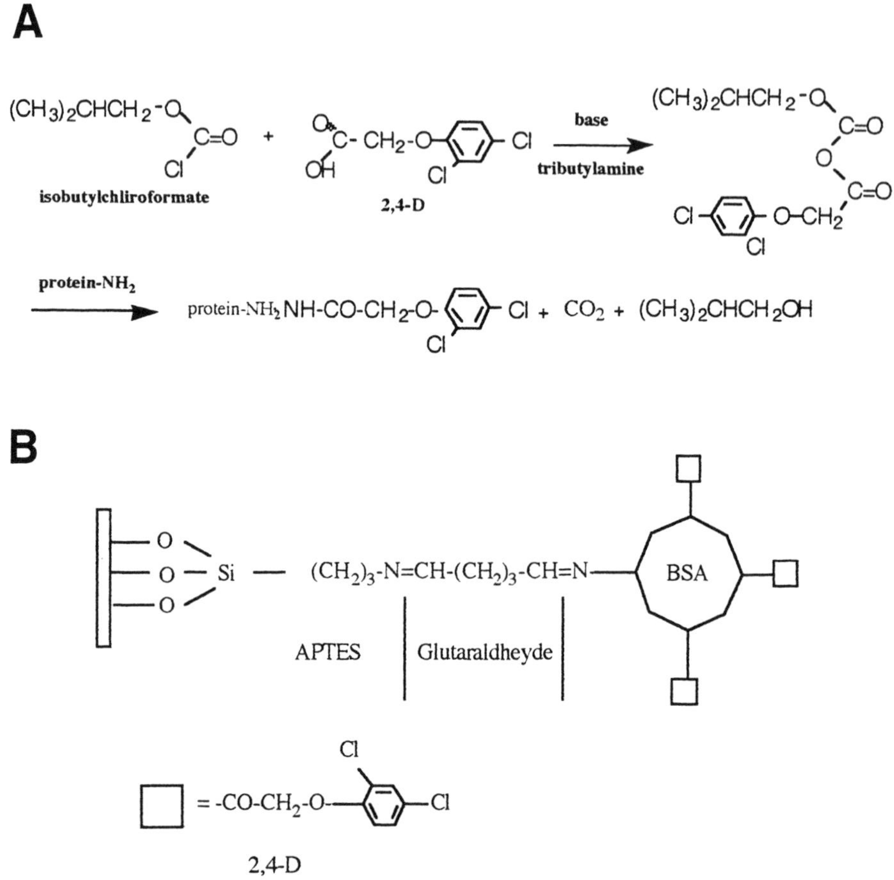

A

B

Scheme 1. **(A)** Chemical modification of the herbicide 2,4-D. **(B)** Surface modification for the coupling of activated 2,4-D.

4. To a new surface with the immobilized 2,4-D, add 500 μL of 10 mM PBS; remove the buffer.
5. Add 500 μL of 1 mg/mL solution of anti-2,4-D antibody from clone 1. Take a different crystal and repeat **steps 4** and **5** using clone 2, instead of clone 1.

No values for the binding constant are given here. For their calculation more experiments are required (*see* **Subheading 3.2.4.**).

3.2.2.3. Calibration Curve for the Analyte

The cell is used in the flow configuration. In the indirect assay the crystals are incubated in the presence of a limited amount of anti-2,4-D antibodies (10 μg/mL) and different concentrations of 2,4-D (sample) for a certain length of time (20 min for each concentration).

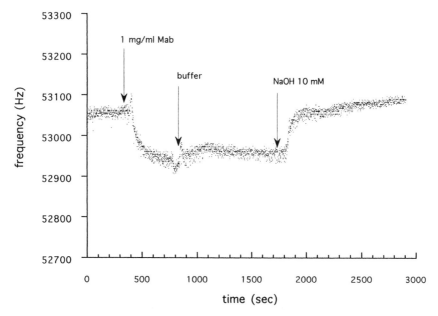

Fig. 6. Real-time monitoring of the interaction between the immobilized 2,4-D and the antibody free in solution. The formation of the immunocomplex is described by changes in the frequency. The surface capacity is tested with 1 mg/mL antibody solution (clone 1). A frequency shift of 113 Hz is found. When the antibody solution is replaced by the buffer, the frequency shift changed very little, indicating that the binding between the antigen immobilized and the antibody is stable and only a regenerating agent, such as 10 mM NaOH, dissociates this binding.

1. Connect the tube with the peristaltic pump and the inlet and the outlet of the cell. Set the flow rate at 75 μL/min.
2. Equilibrate the surface with 10 mM PBS. Put into a vial 1 mL of a solution containing 10 μg/mL anti-2,4-D antibody and a known concentration of antigen (2,4-D).
3. Mix the solution containing the sample with the antibodies and let it flow over the surface. Replace the antibody–antigen solution with the buffer after the interaction.
4. Take the frequency value after the washing.
5. Repeat **steps 2–4** using each time a different crystal and a different antigen concentration. The competition between free and bound 2,4-D for the limited amount of IgG binding sites occurs and the resulting frequency decrease is indirectly proportional to the concentration of free pesticide. No matrix effect has been observed with tap water when the analysis is performed in flow mode. The correlation between the frequency shift and the amount of analyte is good. The analysis time is 30 min for each measurement.

3.2.3. Application 3: Listeria Cells Detection

Bacterial detection is important for clinical, environmental, and food analysis. *L. monocyogenes* is a food pathogen.

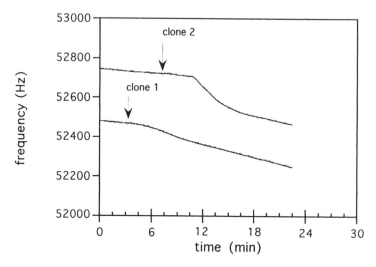

Fig. 7. Binding of two different clones (clones 1 and 2) to the crystal surface. Clone 2 seems to bind the crystal more rapidly than clone 1, giving a frequency shift of 269 and 313 Hz, respectively, indicating a higher affinity for the surface of the first clone.

Precoated crystals with protein A and protein G are used. The method used is a displacement assay (**Fig. 8**).

A static configuration of the cell is used. Twenty microliters of *Listeria* are added and the cell is incubated at 37°C for 1 h. The surface is washed three times with 50 µL DPBS, then 500 µL of DPBS is added and allowed to equilibrate with the surface. The frequency is recorded.

3.2.3.1. DISPLACEMENT ASSAY PROCEDURE
AND CALIBRATION CURVES FOR *L. MONOCYTOGENES*

In the displacement assay, the antigen *Listeria* is immobilized on the crystal surface *(14)*. The addition of the related antibody results in the formation of the immunocomplex with an increase in mass and a corresponding frequency decrease (*see* **Figs. 8** and **9**). The displacement of antibody from the surface is directly related to the concentration of the analyte (*Listeria*) in the sample. To immobilize *Listeria* cells on the crystal surface the following steps are performed:

1. Add to the protein A- or protein G-coated surface 5 µL of 1×10^9 heat-killed *Listeria* cells/mL solution in DPBS. Incubate 1 h at 37°C, wash three times with 500 µL PBS, and measure the frequency.
2. Aspirate the buffer and take 1 mg/mL solution of anti-*Listeria* antibody (from Kirkegaard and Perry) and deposit 30 µL on the crystal. As the binding proceeds, a decrease in frequency is observed until the frequency is stabilized.
3. Replace the antibody solution with 100 µL DPBS to remove the unbound antibody.

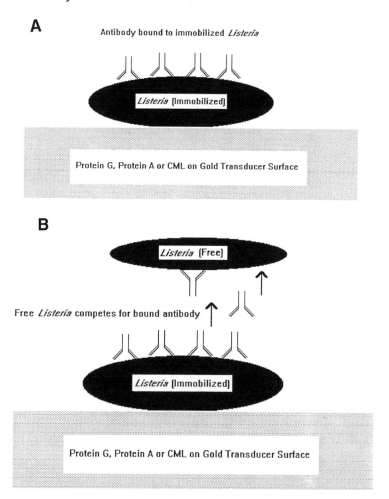

Fig. 8. Scheme of the displacement assay. The antigen *Listeria* is immobilized on the crystal surface (**A**). The addition of the related antibody results in the formation of the immunocomplex with an increase in mass and a corresponding frequency decrease. The *Listeria* cells present in the added sample compete for the antibody site previously bound to the immobilized antigen and cause their displacement with a resulting increase in frequency (**B**).

4. Aspirate the buffer and subsequently add 30 μL of samples containing increasing amounts of *Listeria*.
5. Add a solution of 1×10^7 cells *Vibrio cholerae* Ogawa (kindly supplied by J. Mekalanos, Harvard Medical School, Dept. of Microbiology and Molecular Genetics, Boston, MA) to study the crossreactivity response of the sensor to other microorganisms.

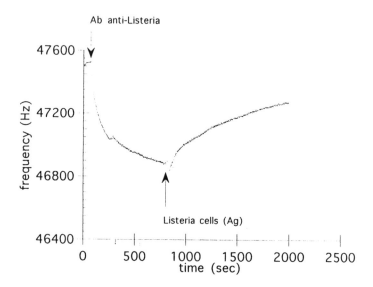

Fig. 9. Real-time monitoring of the antibody displacement when *Listeria* cells are present in solution. The antigen *Listeria* is immobilized on the crystal surface. The addition of the related antibody results in the formation of the immunocomplex and a frequency decrease is observed. After the addition of *Listeria* cells free in the sample, the frequency increases for the antibody displacement. The displacement of antibody from the surface is directly related to the concentration of the analyte (*Listeria*) in the sample.

With the addition of specific antigen, a frequency increase is observed depicting the free *Listeria* cells' displacement of the antibody competing with bound antigen. **Figure 10** shows the results obtained with protein A- and G-coated crystal.

3.2.3.2. *L. monocytogenes* Detection in Milk

1. Add 900 µL of milk (2% fat) to a prelabeled competition crystal mounted in the flow cell. (Crystals are prepared as described in **Subheading 3.2.3.**).
2. Equilibrate the cell for 15 min.
3. Add 10 µL of the nonspecific antigen *Serratia* (6×10^6 cells) to the cell, mix, and record the reaction rate (slope) using PZTools.
4. Make a second addition of *Listeria* (3.19×10^6 cells) to the flow cell solution and mix.

The slope of this data is calculated (using PZTools) and compared with the nonspecific slope results. The values are reported in **Table 1**.

In addition to concentration measurements in real-time, a big advantage of such a device is the possibility of directly following the reaction while it is occurring.

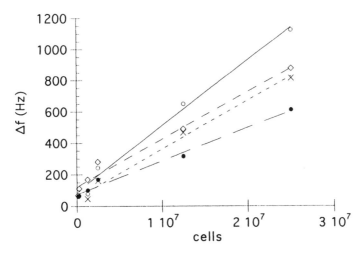

Fig. 10. Standard curves for *Listeria* when two different crystal coatings (protein A and protein G) are used.

Table 1
Slopes Generated by Addition
of Either *Serratia* (Nonspecific)
or *Listeria* in 2% Milk and Evaluated with PZTools

Organism	Cell number/mL[a]	Slope[b]
Serratia	6×10^6	1.257
Listeria	3.19×10^6	4.651
Listeria	6.3×10^6	4.665

[a]10 μL of cell are added to the protein G precoated surface. First is added the nonspecific antigen (*Serratia*, first line) followed by subsequent addition of specific antigen (*Listeria*, second line). A separate addition of specific antigen (*Listeria*, third line) is also reported.
[b]The slope is taken after antigen addition.

3.2.4. Application 4: Measuring Affinity and Rate Constants

Quartz crystal microbalance (QCM) can be used to study the affinity reaction between two interacting molecules, i.e., antibody and antigen *(9)*. Characterization of the affinities and rates of these interactions is very important in many fields. Although conventional assay methods achieve high sensitivity using radioactive, fluorescent, or enzyme labels, they often require a time-consuming incubation and separation step; in general, the reaction is allowed to reach the equilibrium, and free and bound reactants are separated and quantified. Often these methods tell about the equilibrium constant value but they do not tell anything about the time necessary to reach the equilibrium (association rate constant) and also about the rate of complex dissociation (dissociation rate constant). These constants are available in PZTools.

3.2.4.1. THEORY

3.2.4.1.1. Equilibrium Constant

The equilibrium constant K characterizes the molecular interactions. Mixing in solution two interacting molecules, A and B, they will form a complex until equilibrium is reached:

$$A + B \rightleftarrows AB \tag{2}$$

The position of the equilibrium is dependent on the concentrations of A, B, and AB (law of mass action) and can be described by the equilibrium constant K. Depending on the direction of the equilibrium, K can be expressed as an association equilibrium constant K_A (M^{-1}) or a dissociation equilibrium constant K_D (M). These constants are the inverse of each other:

$$[AB]/[A][B] = K_A, \text{ or } [A][B]/[AB] = K_D \tag{3}$$

3.2.4.1.2. Association and Dissociation Rate Constants

The affinity constant does not completely describe the reaction; it is also important to know the association and dissociation rates. The equilibrium is a dynamic event; complex formation and dissociation continuously occurs. The K does not contain any information about the time required to reach equilibrium. The rates of association and dissociation are described by association and dissociation rate constants:
Association:

$$A + B \rightarrow AB \tag{4}$$

As the interaction proceeds, the AB concentration increases and we can then write:

$$(d[AB]/dt) = k_a[A][B] \tag{5}$$

where k_a $(M^{-1} \cdot s^{-1})$ is the association rate constant. This constant indicates how fast the concentration of AB complexes increases when the interactants A and B are present at a concentration of [A] and [B].
Dissociation:

$$AB \rightarrow A + B \tag{6}$$

The decrease in concentration of AB complexes over time is then:

$$(d[AB]/dt) = k_d[A][B] - k_d[AB] \tag{7}$$

where k_d (s^{-1}) is the dissociation rate constant, which indicates the AB complex dissociated per second.

The amount of complex AB formed over time is the the sum of the association and dissociation expressions:

$$(d[AB]/dt) = k_a[A][B] - k_d[AB] \tag{8}$$

At the equilibrium the sum is 0, and the ratio between the k_a / k_d is K_A and the inverse K_D.

3.2.4.2. PZ Crystal Approach (All on PZTools)

The complex formation can be monitored applying the rate equation, which can be done using the frequency changes on a piezoelectric crystal. Considering the Sauerbrey equation ($\Delta f = -2.3 \times 10^{-6} F^2 \Delta m/A$ [the decrease of frequency f is directly proportional to the attached mass]) and using f_m as the frequency change after a complete saturation of the surface of the crystal with antibodies, the concentration of the free antigen B, [B], is proportional to $(f_m - f)$ and the concentration of the complex AB, [AB], to f.

Equation 1 can thus be expressed as:

$$df/dt = k_a c(f_m - f) - k_d f \tag{9}$$

where c is the concentration of the free antibody A, held constant in a continuously flowing solution.

For the determination of the constants k_a and k_d from the experimental f vs t curves, the derivatives df/dt must be plotted versus the corresponding frequency changes f. A straight line characterized by a slope SL and intercept INT should be obtained according to **Eq. 2**. The parameters SL and INT are related to the kinetic constants:

$$SL = -(k_a c + k_d) \tag{10}$$

$$INT = k_a f_m c \tag{11}$$

By measuring the binding curves (f vs t) determined for several concentrations c, all of the desired parameters k_a, k_d, and f_m could be obtained. The equilibrium constant K_A for the AB can be obtained as a ratio: $K_A = k_a/k_d$.

The antibody–antigen interaction at solid–liquid interfaces is often limited by diffusion kinetics, especially for immobilized antigen formats (owing to significant diffusion constants of large-mol-wt antibodies to the surface). It has been noted that diffusion plays a significant part when high concentrations of antigens (or binding sites) are immobilized on the surface. Even in the cases in which the reaction is not diffusion-limited, measured forward and reverse reaction rates are typically lower for surface reactions than for solution reactions *(15)*.

Table 2
Kinetic and Equilibrium Constants Characterizing Affinity Interactions of the Derived Antibodies with 2,4-Dichlorophenoxyacetic Acid Conjugated with Albumin Immobilized on the Surface of Piezoelectric Crystals

MAb	K_a(mol/L/s)	K_d(10^{-3} s)	K_A(10^7/mol/L)
Clone 1	4540	0.789	0.575
Clone 2	9780	1.87	0.523
Clone 3	9010	1.10	0.819
Clone 4	12,900	0.913	1.42
Clone 5	12,300	1.74	0.707

3.2.4.3. AFFINITY AND RATE CONSTANTS OF DIFFERENT MAb CLONES

An example here is a study on the characterization of monoclonal antibodies (MAb) prepared against the 2,4-D herbicide in the laboratory of M. Franek. The cell is used in the flow mode.

1. Connect the tube with the peristaltic pump and also with the inlet of the cell. Set the flow rate to 75 μL/min.
2. Equilibrate the surface with 10 m*M* PBS and then inject a solution containing a known amount of anti-2,4-D antibody (starting from a concentration of 10 up to 100 μg/mL).
3. Replace the antibody–antigen solution with the buffer and regenerate the surface by injecting 10 m*M* NaOH for 10 min.
4. Repeat from **steps 2** to **4** using a different crystal antibody concentration each time.

The regeneration procedure is quite reproducible for 10 consecutive cycles of binding. The time–frequency curves resulting from the binding of MAb to 2,4-D immobilized on the surface of QCM are used for calculations of the association and kinetic constants of the MAbs. A plot of *df/dt* vs frequency for different concentrations of antibodies is obtained. The dependence of slope (*SL*) and intercept (*INT*) on MAb concentration is calculated for five MAb clones and the calculated constant values are reported in **Table 2**.

4. Notes

1. The piezoelectric device is a suitable technique for real-time monitoring of biospecific interactions, such as antigen–antibody reaction, receptor–ligand binding, and so forth, without the use of any label. These are great advantages because it is possible to detect molecules without any prior purification steps or any reagent's conjugation to labels as shown in **Subheadings 3.2.1.–3.2.3.**
2. It is important to be aware of some problems related to the crystals, such as the fact that the crystals are fragile and they should be handled with care.

3. Some problems of nonspecific binding could be found when not only the specific molecules are bound to the surface. This nonspecific binding has been studied for the three coatings (avidin, protein A or protein G, and silanization) and resulted in 30% of unspecific reaction with an anti-BSA antibody solution on the avidin-biotin–antibody-coated surface (*see* **Subheading 3.2.1.**). Nonspecific binding is at the same level for protein A or protein G (**Subheading 3.2.3.**) and is around 15% for the silanized surfaces (**Subheading 3.2.2.**). A blank control is always recommended.

4. It is possible to reuse the crystal by regenerating the surface using low pH solutions to dissociate the antigen–antibody binding. Either 0.1 *M* glycine, pH 2.5 (**Subheading 3.2.1.**) or 10 m*M* NaOH are suitable for this purpose (**Subheading 3.2.4.**). The regenerating agent is not suitable for the protein A- or protein G-coating since in this case the protein A- or protein G-binding with the antibody, or the antibody itself will be dissociated.

5. The sensitivity of the system is around ppb for small molecules (**Subheading 3.2.2.**) using an indirect assay, ng/mL for molecules like ricine, μg/mL for antibody solutions, and 10^5 cells for bacterial detection (**Subheading 3.2.3.**).

6. The analysis time is about 5–10 min for each sample.

7. Kossliger et al. *(18)* have described simultaneous measurements with the QCM and surface plasmon resonance (SPR), the technique used in the Pharmacia BIAcore instrument, using a specially designed flowthrough cell and a flow-injection analysis (FIA) technique. These authors compared the influence of sample properties, as well as surface properties, in a study of unspecific binding of BSA to gold surfaces, specific binding between BSA and MAbs, and also regeneration and serum experiments with human immunodeficiency virus antigens. The sensitivity of both methods is nearly the same, and for the reaction of antibodies with antigens an angle change of 1 millidegree can be assigned a frequency change of 8 Hz, which is about 8 ng/cm^2. They concluded that the two systems were equivalent in sensitivity and cross-selectivity.

References

1. Ward, M. D. and Buttry, D. A. (1990) In situ interfacial mass detection with piezoelectric transducer. *Science* **249,** 1000.
2. Luong, J. H. T. and Guilbault, G. G. (1991) Analytical applications of piezoelectric crystal biosensor, in *Biosensor Principles and Applications* (Blum, L. J. and Coulet, P. R., eds.), Marcel Dekker, New York, p. 107.
3. Sauerbrey, G. Z. (1959) Use of quartz vibrator for weighing thin films on a microbalance. *Z. Phys.* **115,** 205.
4. Dunham, G. C., Benson, N. H., Petelenz, D., and Janata, J. (1995) Dual quartz crystal microbalance. *Anal. Chem.* **67,** 267.
5. Kanasawa, K. K. and Gordon, J. G. (1985) Frequency of a quartz microbalance in contact with liquid. *Anal. Chem.* **57,** 1771.
6. Kurosawa, S., Tawara, E., Kamo, N., and Kobatake, Y. (1990) Oscillating frequency of piezoelectric quartz crystal in solutions. *Anal. Chim. Acta* **230,** 41.

7. Bruckenstein, S. and Shay, M. (1994) Dual quartz crystal oscillator circuit-minimizing effect due to liquid viscosity, density, and temperature. *Anal. Chem.* **66,** 1847.
8. Thompson, M., Kipling, A. L., Duncan-Hewitt, W. C., Rajakovic, L. V., and Cavic-Vlasak, B. A. (1991) Thickness-shear-mode acoustic wave sensor in the liquid phase: a review. *Analyst* **116,** 881.
9. Thompson, M., Artur, C. L., and Dhaliwal, G. K. (1986) Liquid phase piezoelectric and acoustic transmission studies on interfacial immunochemistry. *Anal. Chem.* **58,** 1206.
10. Ghourchian, M. O. and Kamo, N. (1995) New detection cell for piezoelectric quartz crystal: frequency changes strictly follow Bruckenstein and Shay's equation in very dilute non-electrolyte aqueous solution. *Analyst* **120,** 2737.
11. Weetall, H. H., ed. (1995) *Immobilized Enzymes, Antigens, Antibodies, and Peptides—Preparation and Characterization,* vol. 1. Marcel Dekker, New York.
12. Davis, K. A. and Leary, T. (1989) Continuous liquid phase piezoelectric biosensor for kinetic immunoassay. *Anal. Chem.* **61,** 1227.
13. Starzi, S., Santori, T., Minunni, M., and Mascini, M. (1998) Surface modifications for the development of piezoimmunosensors. *Bios. Biol.,* in press.
14. Minunni, M., Mascini, M., Carter, R. M., Jacobs, M. B., Lubrano, G. J., and Guilbault, G. G. (1996) A quartz crystal microbalance displacement assay for listeria monocytogenes. *Anal. Chim. Acta* **325,** 169.
15. Nieba, L., Krebber, A., and Plükthun, A. (1996) Competition BIAcore for measuring true affinities: large differences from values determined from binding kinetic. *Anal. Biochem.* **234,** 155–165.
16. Kosslinger, C., Drost, S., Aberl, A., Wolf, H., Brink, G., Stangmaler, A., and Sackmann, E. (1994) Comparison of the QCM and SPR method for surface studies and immunological applications. International Conference on Chemical Sensors, Rome, Italy.

5

Immunobiosensors Based on Evanescent Wave Excitation

Randy M. Wadkins and Frances S. Ligler

1. Introduction

Evanescent wave immunobiosensors use antibodies immobilized at the surface of a waveguide to form a sensing device. Fluorescence-based evanescent wave biosensors form a fluorescent complex at the surface of the waveguide when the antigen of interest is present (e.g., a sandwich assay or direct binding of a fluorescent analyte) or alternatively, the nonfluorescent analyte may compete with a fluorescent analog of the analyte for binding, as in a competitive immunoassay format. Light propagating through the waveguide extends a short distance into the surrounding medium and excites the immobilized fluorophore. The waveguide also collects the emitted fluorescence and carries this light to a photodetector.

The use of evanescent wave excitation of fluorescence has many advantages over other types of biosensors for the analysis of complex or "dirty" solutions, such as whole blood, serum, urine, seawater, and groundwater. First, these types of samples contain proteins and other molecules that are notoriously sticky and generate high background signals when measurements are conducted using mass-sensitive devices, i.e., surface plasmon resonance systems, acoustic wave sensors, piezoelectric sensors, and interferometric approaches. Since the evanescent measurements of fluorescence at longer wavelengths (e.g., 650 nm) only detect the exogenously labeled molecules localized at the fiber surface, the interference from natural compounds (which fluoresce at shorter wavelengths) or from fluorophores in the bulk solution is minimal. Second, the evanescent wave measurement is also better suited to relatively opaque samples than cuvet-based measurements of fluorescence or absorption (1) because it is not necessary to propagate light through the solution to make a measurement.

From: *Methods in Biotechnology, Vol. 7: Affinity Biosensors: Techniques and Protocols*
Edited by: K. R. Rogers and A. Mulchandani © Humana Press Inc., Totowa, NJ

Only the region immediately next to the fiber surface needs to be illuminated. Finally, the "immuno" function of the sensor facilitates detection of antigens at low concentrations that would be difficult to detect via standard spectroscopic methods, since the antigen is concentrated by the antibodies at the surface of the biosensor.

Several excellent review articles are available on the theory of waveguides and production of evanescent wave excitation *(2–4)*, and therefore we will only briefly cover the topic. When a ray of light propagating through a medium strikes an interface with another medium having a different refractive index (i.e., glass and air, or glass and water), it will be totally internally reflected if the angle of incidence is below a critical angle, defined by $\sin^{-1}(n_1/n_2)$, where n_1 is the refractive index of glass and n_2 is the refractive index of water. These conditions are usually met with an optical fiber. Light propagating through a glass fiber will be contained within the fiber when the interface is air or water, and therefore the fiber will act as a waveguide.

The intensity of light striking the waveguide interface does not fall immediately to zero at the interface. Because of the wave properties of light, a portion of the light intensity will leave the fiber, travel a short distance down the length of the fiber, and reenter it. The intensity of the light leaving the glass decays exponentially away from the interface, and the distance at which the intensity of the light decays to $1/e$ of its interface intensity is known as the penetration distance. The penetration distance is on the order of 100 nm from the interface, and the component of light within this distance is known as the evanescent wave. The evanescent wave is of sufficient intensity to excite fluorophores within the penetration distance, but is too low in intensity to excite those found much farther away from the interface.

In general, any object that acts as a waveguide can be made into an evanescent wave biosensor *(5)*. Hirshfeld and Block *(6)* first proposed the use of optical fibers as waveguides for conducting fluorescent immunoassays. These fiberoptic biosensors are constructed by immobilizing antibodies on the surface of the core of the fiber, after removal of the cladding and buffer surrounding the core. Fiberoptic biosensors have been used to analyze a variety of aqueous samples for proteins *(7,8)*, bacteria *(8)*, or small molecules *(9–11)*. Other laboratories have developed evanescent wave biosensors based on planar waveguides (e.g., **refs. *12,13***).

To construct a fiberoptic biosensor, commercial optical fibers are stripped of their cladding and buffer a short distance (12 cm) from the distal end. However, immersion of the unclad fiber into an aqueous solution results in a mode mismatch between the immersed stripped core and the clad portion of the fiber. Fluorescence enters the fiber in the higher order modes, and light in higher order modes is preferentially lost from the fiber at the interface between the clad and stripped regions *(14,15)*. To enhance efficiency of fluorescence detection, the exposed core is tapered to a geometry that reduces this mode mismatch. The "combination" taper geometry, described in **Subheading 3.2.**, improves signal detection more than 100 times that of an unetched fiber *(7,14)*.

Antibodies are immobilized on the fiber core by first coating it with a thiol-containing silane, followed by a heterobifunctional crosslinker that reacts with both the exposed thiol groups of the silane and the terminal amino groups of the antibody *(16)*. Alternatively, protein G or avidin can be immobilized directly to the fiber probes, and the antibodies bound to the fiber via the Fc region of the immunoglobulin or biotinylated site on the antibody, respectively.

The fibers are then mounted into a chamber, and samples are flowed over the biosensor. Detection of antigens in the sample may be done as a direct assay *(7)*, a sandwich assay (described in **Subheading 3.3.2.**), or as a competitive immunoassay *(9–11)*. This latter method is particularly suited for antigens too small to bind to more than one antibody at a time.

The simplest optical design for fiberoptic biosensor measurements is given below. However, a portable device based on this design is now commercially available (Analyte 2000, Research International, Woodlinville, WA) and has the ability to interrogate four fiber probes simultaneously. Fibers with preattached connectors that can be tapered for use with either the Analyte 2000 or the benchtop system described in **Subheading 2.1.** are also available commercially (Research International).

2. Materials
2.1. Basic Biosensor Construction (Refer to Fig. 1)

1. Optical breadboard with assorted posts and positioning equipment (available from Melles Griot [Auburn, MA], Newport [Irvine, CA], Edmund Scientific [Barrington, NJ], and so forth).
2. A 5-mW 635-nm diode laser (LaserMax, Rochester, NY).
3. Mirror with mount (Newport).
4. Two 1-in. lenses with mounts (Newport).
5. A 645-nm dichroic beamsplitter (Omega, Ontario, CA).
6. A 665-nm long-pass filter (Schott Glass, Yonkers, NY).
7. Lock-in amplifier and mechanical chopper (Stanford Instruments models SR510 and SR540, respectively).
8. Photodiode (EG&G model SGD-100A). The photodiode may be used in reverse bias mode, requiring a 10-ΩW resistor and a 9-V battery.
9. Thin steel or aluminum plate (~5 × ~7.5 cm) to act as a fiber mounting bracket.
10. Female SMA- or ST-type connector.
11. Epoxy cement (5-min curing).
12. Optical fiber cleaver.

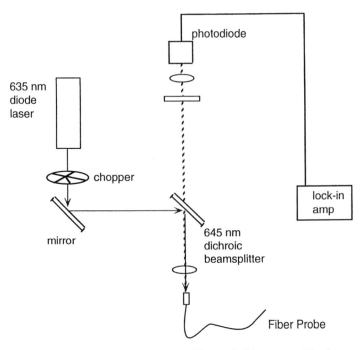

Fig. 1. The configuration for setting up a fiberoptic biosensor. The laser excitation is focused into the fiber using a mirror, a dichroic beam splitter, and a focusing lens. The returning fluorescence passes through the dichroic and is focused onto a photodiode detector that is synchronizedd to a mechanical chopper via a lock-in amplifier.

2.2. Fiber Preparation

1. Male SMA- or ST-type connector.
2. Plastic-clad, silica-core optical fibers (Quartz et Silice [Tuckerton, DE] or Research International).
3. Fiber polisher or fine sandpaper with 12-, 6-, 3-, 1-, and 0.3-μm grains.
4. Concentrated hydrofluoric acid (J.T. Baker, Phillisburg, NJ).
5. Programmable stepper-motor or worm-drive system (Parker Hannifin Corp., Cleveland, OH).
6. Rack for mounting fibers into stepper-motor unit.
7. Polypropylene plastic graduated cylinder.
8. Teflon or HF-resistant (polypropylene) plastic pipets (2-5 mL).
9. Razor blades.

2.3. Antibody Immobilization

1. 100 mL Methanol: conc. HCl (50:50) solution.
2. 100 mL Concentrated sulfuric acid.
3. Distilled water for rinsing.

4. 100 mL Boiling distilled water.
5. 3-Mercaptopropyltrimethoxysilane (Fluka, Buchs, Switzerland).
6. Toluene.
7. Nitrogen tank (optional).
8. *N*-Succinimidyl-4-maleimidobutyrate (Fluka).
9. Dimethylformamide.
10. Ethanol.
11. Antibody solution (0.05 mg/mL) in phosphate-buffered saline (PBS, 2–5 mL for every five fibers).
12. PBS, pH 7.4.
13. Disposable 2- or 5-mL glass pipets, flame sealed.

2.4. Immunoassay

1. 200-µL Capillary tubes (Fisher, Pittsburgh, PA).
2. Silicone tubing, 1/16-in. od, 1/8-od (Cole-Parmer [Mernon Hills, IL] #06411-62).
3. Nylon "T" connectors (Bio-Rad [Hercules, CA] #732-8302).
4. Disposable plastic 1-mL syringes.
5. Hot glue gun and glue (hardware store).
6. Blocking buffer: 2 mg/mL bovine serum albumin, 2 mg/mL casein, 0.1% Triton X-100 in PBS.
7. Cy5-labeled secondary antibodies (Jackson Immunoresearch, West Grove, PA).

3. Methods (*see* Notes 1–3)

3.1. Basic Biosensor Construction (Refer to *Fig. 1; If Using the RI Device, Skip to* Subheading 3.2.)

1. Drill a hole into one end of the metal plate, large enough such that the female connecter just fits into the hole. Use the 5-min epoxy to seal the connector into the plate.
2. Drill a hole in the plate on the edge distant from the connector, large enough for the optical bench screws. This will allow mounting of the holder onto the optical bench.
3. Mount the fiber into the fiber holder using the connectors (*see* **Note 1**).
4. Configure the optical breadboard as shown in **Fig. 1**, using the mirror and dichroic filter to focus the excitation laser into the fiber.
5. Align the photodiode with the returning fluorescence (*see* **Notes 2** and **3**).
6. Adjust the phase of the lock-in chopper to obtain the maximum signal from the returning fluorescence.

3.2. Fiber Preparation

3.2.1. Attaching a Connector to the Fiber

1. Using a razor blade and a fiber cleaver, cut an appropriate length of fiber (usually approx 1 m). Trim a small amount of the cladding and buffer from the core (~0.5 cm) at one end of the fiber. Using the cleaver, cut the partially stripped core at the end, making as blunt a cut as possible (*see* **Fig. 2**).

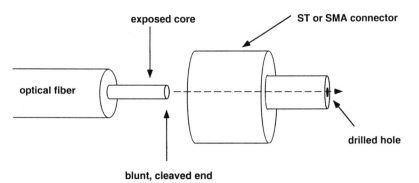

Fig. 2. Fitting of the connector onto the optical fiber. The fiber is cleaved and stripped a short distance from one end. A small hole, just larger than the fiber diameter, is drilled through an SMA- or ST-type connector. The fiber is placed through the connector and sealed in place with epoxy. The end of the fiber is made flush with the connector by polishing the with fine sandpaper.

2. Drill a hole in the male connector just large enough to insert one end of the optical fiber.
3. At the partially stripped end of the fiber, coat the fiber liberally with freshly mixed epoxy. Insert the partially stripped end of the fiber through the connector body until it just emerges from the hole in the connector. Fill the body of the connector with mixed epoxy and allow it to dry fully (*see* **Fig. 2**).
4. Using the sandpaper in the order 12-, 6-, 3-, 1-, and finally 0.3-μm grains, polish the connector fiber end attached to the connector until the fiber core is even with the connector end and the core appears flat and smooth. This proximal end will then be mounted into the optical system.

3.2.2. Predip Procedure

1. Using a razor blade, strip 13 cm of the plastic buffer and plastic cladding from the distal end of the fiber.
2. Using laboratory tape, group fibers into bundles of five. Make certain tape is well away from the stripped area of the fibers.
3. In a fume hood, place concentration HF into a plastic graduated cylinder or clamped tygon tube.
4. Place one fiber bundle into the HF for 1 min. Time is critical, so as to not significantly etch the fibers.
5. Rinse the fiber bundle in water to remove HF. Remove the loosened residues of cladding and buffer from the stripped area.
6. Store fibers until ready to taper.

3.2.3. Tapering Procedure (see **Notes 4** and **5**)

1. Trim stripped area of fibers to 12.5 cm using a razor blade. Mount fibers into the dipper assembly (*see* **Fig. 3** and **Note 4**).

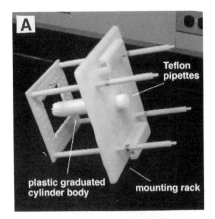

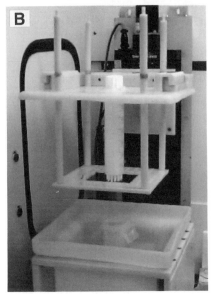

Fig. 3. Tapering fibers with HF. **(A)** Teflon pipets (2–5 mL) are placed into the body of a plastic graduated cylinder (500 mL) and sealed in place using epoxy. The graduated cylinder is then sealed onto a plastic or plexiglass sheet for mounting onto the stepper-motor. Stripped and cleaned fibers are placed through individual pipets and held in place with rubber stoppers. **(B)** The mounted rack is subsequently mounted onto the stepper-motor, and positioned to lower the fibers into a graduated cylinder filled with concentrated HF.

2. Using the dipper, lower the fibers into a plastic graduated cylinder filled with concentration HF. The rate of lowering will depend on the initial and final fiber diameter desired, as well as the shape of the taper. For a 600-μm core fiber, we

immediately immerse the stripped core 2 cm, followed by a rate of 3 cm/min for 10 of the remaining 10.5 cm. This produces a combination taper with a ~180-μm-end radius (*see* **Note 5**).
3. Immediately immerse the fibers in water following the tapering. Rinse twice with water.
4. The HF will need to be replaced after approx 25 dips.

3.2.4. Cleaning Procedure

1. Prepare 50:50 methanol:HCl (concentration) in a 100-mL graduated cylinder. Place fibers in solution for 30–60 min.
2. Rinse with distilled water at least twice.
3. Place fibers in concentrated sulfuric acid (in 100-mL graduated cylinder) for 30–60 min.
4. Rinse with water at least three times.
5. Pour boiling water into 100-mL graduated cylinder and immerse fibers for 15–20 min (it is not necessary to keep water boiling).
6. Splay and air-dry fibers. Go immediately to silanization (**Subheading 3.2.5.**).

3.2.5. Silanization Procedure (see **Notes 6** and **7**)

1. Prepare a 2% solution of 3-mercaptopropyltrimethoxysilane in toluene (1 mL silane to 50-mL toluene). Fill a 50 mL graduated cylinder with the solution (*see* **Note 6**). Cap with Teflon tape.
2. Purge the silane/toluene solution with nitrogen to inhibit oxidation of thiol groups (optional).
3. Place the fibers in the silane solution. Optionally, continue to purge with nitrogen during the immersion period. Let fibers stand for 30 min to 2 h.
4. Rinse fibers twice with pure toluene and allow them to air dry (*see* **Note 7**). Proceed immediately to crosslinking step (**Subheading 3.2.6.**).

3.2.6. Crosslinking Antibodies

1. Prepare a 2 mM solution of N-succinimidyl-4-maleimidobutyrate in ethanol (14 mg in 25 mL ethanol). Dissolve the crosslinker in a few drops of dimethylformamide before mixing it with the ethanol.
2. Place fibers in the crosslinker solution in a 25-mL graduated cylinder for 1 h. Cover the cylinder with Parafilm to prevent evaporation.
3. Rinse fibers three times with PBS.

3.2.7. Protein Immobilization

1. Prepare antibody solution in PBS at 0.05 mg/mL.
2. Place antibody solution into flame-sealed 2 or 5 mL disposable pipets. Place five fibers in each pipet and hold them in place by inserting a rubber stopper alongside the fibers in the open pipet end. Incubate 1 h at room temperature (25°C).
3. Remove fibers and store in a flame-sealed pipet containing only PBS. The antibody retains its activity for several weeks under these conditions.

3.3. Fluorescence Immunoassay

3.3.1. Flow Chamber

1. Fit a 0.5-cm length of silicone tubing over each end of a capillary tube.
2. Insert the long axis of a T-connector into the silicone tubing at both ends of the capillary tube such that the base of each T is pointing upward.
3. Attach a longer length of silicone tubing to one of the T-connector bases. The length of tubing should be long enough to reach a waste container. This will be referred to as the distal end of the chamber.
4. Insert a fiber through the proximal T-connector, through the capillary tube, and allow it to emerge slightly from the distal T-connector (prewetting the tube with PBS will facilitate this procedure).
5. Use hot glue to seal the fiber into the tube at both the proximal and distal T-connector. Seal the end in-line with the fiber, not the perpendicular end.
6. Mount the flow chamber onto a supporting surface (the edge of a lab bench works well).
7. Fit a 1-cm length of silicone tubing over the end of a 1-mL syringe (without needle). The syringe will be used to deliver buffer and samples into the flow chamber.

3.3.2. Sample Measurement

1. Connect the proximal end of the fiber to the biosensor. Always block laser excitation when not measuring antibody signals to avoid photobleaching of sample.
2. Using the 1-mL syringe, wash the fiber twice with blocking buffer. Allow the buffer to sit for 10–20 min to block nonspecific protein binding sites on the fiber (this and subsequent steps are performed at ambient temperature).
3. Introduce the sample containing the antigen to the chamber. Incubate 5 min.
4. Wash once with blocking buffer and record signal.
5. Introduce Cy5-labeled secondary antibody solution (5–10 µg/mL in blocking buffer) to the chamber. Incubate 1–5 min. Record signal as voltage increase at 30-s intervals (the data collection is automatically done using the Research International Analyte 2000).
6. Repeat **steps 3–5** for all samples.
7. A reference signal for each fiber can be generated by using a concentrated antigen solution, and repeating **steps 3–5**. Typically, a 1-µg/mL solution of antigen is used to generate a signal for the completely saturated fiber.

4. Notes

4.1. Basic Biosensor Construction

1. The mounting bracket will depend on the fiber coupler used. A typical mounting bracket might be a female SMA connector mounted into a steel plate. The fiber can then be secured into the male SMA connector by drilling a hole, just slightly larger than the fiber diameter, into the connector and sealing the fiber in place using epoxy. The end of the mounted fiber must then be polished to enhance light

coupling and reduce reflected light. The Analyte 2000 device uses an ST-type connector.

2. The photodiode may need to be operated in the reverse bias mode to ensure a linear response over a wide range of signal intensities. The manufacturer of the photodiode should provide a schematic of the simple circuit (which uses a battery and a resistor) that will work best for the photodiode.

3. An effective way to align the photodiode is to use either a longer wavelength laser (e.g., 560 nm) focused onto the distal end of the fiber, or a concentrated solution of Cy5-labeled protein attached to the distal end of the fiber.

4.2. Fiber Preparation

4. The fibers should be kept separate during dipping. This can be accomplished by constructing a device for mounting the fibers, with Teflon or plastic pipets of appropriate length glued into a large plastic tube. Such a tube can be made by removing the base of a plastic graduated cylinder. This tube may then be mounted onto a board that can subsequently be mounted into the dipper apparatus (*see* **Fig. 3**).

5. The dipping time will need to be calibrated to produce the desired taper. Use of an optical microscope to measure the diameter of a fiber after a period of immersion can be used to produce a graph of core diameter vs dipping time. The dipper can then be programmed accordingly.

6. The silane should be stored under nitrogen or argon and should only be used for 1–2 mo after opening the bottle.

7. Examine the fibers for the presence of a white film or other particulate matter. The silane will oxidize with time, which produces a material that will coat the fiber. If this occurs, start the procedure again using fresh silane.

Acknowledgments

These procedures are a summary of various techniques that have been worked out by Richard Thompson, Lisa C. Shriver-Lake, Joel P. Golden, and George P. Anderson, with helpful tips from Keeley King and Kristen Breslin. The opinions expressed in this work are those of the authors and do not necessarily represent those of the US Government, Department of Defense, or the US Navy. Wadkins was supported through the National Research Council Associateship Program at the Naval Research Laboratory.

References

1. Golden, J. P., Anderson, G. P., Ogert, R. A., Breslin, K. A., and Ligler, F. S. (1992) An evanescent wave fiber optic biosensor: challenges for real world sensing. *SPIE Proc.* **1796,** 2–8.

2. Thompson, R. B. (1991) Fluorescence-based fiber-optic sensors, in *Topics in Fluorescence Spectroscopy, vol. 2: Principles* (Lakowicz, J. R., ed.), Plenum, New York, pp. 345–365.

3. Thompson, R. B. and Ligler, F. S. (1991) Chemistry and technology of evanescent wave biosensors, in *Biosensors with Fiberoptics* (Wingard, L. B. and Wise, D. L.), Humana, Totowa, NJ, pp. 111–138.
4. Axelrod, D. (1989) Total internal reflection fluorescence microscopy. *Meth. Cell Biol.* **30**, 245–270.
5. Kronick, M. N. and Little, W. A. (1975) A new immunoassay based on fluorescence excitation by internal reflection spectroscopy. *J. Immunol. Meth.* **8**, 235–240.
6. Hirshfield, T. E. and Block, M. J. (1984) Fluorescent immunoassay employing optical fiber in capillary tube. U.S. Patent No. 4,447,546.
7. Ligler, F. S., Golden, J. P., Shriver-Lake, L. C., Ogert, R. A., Wijesuria, D., and Anderson, G. P. (1993) Fiber-optic biosensor for the detection of hazardous materials. *Immunomethods* **3**, 122–127.
8. Shriver-Lake, L. C., Ogert, R. A., and Ligler, F. S. (1993) A fiber-optic evanescent-wave immunosensor for large molecules. *Sensors Actuators B* **11**, 239–243.
9. Shriver-Lake, L. C., Breslin, K. A., Charles, P. T., Conrad, D. W., Golden, J. P., and Ligler, F. S. (1995) Detection of TNT in water using an evanescent wave fiber-optic biosensor. *Anal. Chem.* **34**, 2431–2435.
10. Walczak, I. M., Love, W. F., Cook, T. A., and Slovacek, R. E. (1992) The application of evanescent wave sensing to a high-sensitivity fluoroimmunoassay. *Biosensors Bioelectron.* **7**, 39–48.
11. Oroszlan, P., Thommen, C., Wehrli, M., Duveneck, G., and Ehrat, M. (1993) Automated optical sensing system for biochemical assays: a challenge for ELISA? *Anal. Meth. Instrument.* **1**, 43–51.
12. Christensen, D., Johannson, T., and Petelenz, D. (1994) Biosensor development at the University of Utah. *IEEE Engineer. Med. Biol.* **13**, 388–395.
13. Reichert, W. M., Ives, J. T., Suci, P. A., and Hlady, V. (1987) Excitation of fluorescent emission from solutions at the surface of polymer thin-film waveguides: an integrated optics technique for the sensing of fluorescence at the polymer/solution interface. *Appl. Spec.* **41**, 636–640.
14. Anderson, G. P., Golden, J. P., and Ligler, F. S. (1993) A fiber optic biosensor: combination tapered fibers designed for improved signal acquistion. *Biosensors Bioelectron.* **8**, 249–256.
15. Golden, J. P., Anderson, G. P., Rabbany, S. Y., and Ligler, F. S. (1994) An evanescent wave biosensor—part II: fluorescent signal acquisition from tapered fiber optic probes. *IEEE Trans. Biomed. Engineer.* **41**, 585–591.
16. Bhatia, S. K., Shriver-Lake, L. C., Prior, K. J., Georger, J. H., Calvert, J. M., Bredehorst, R., and Ligler, F. S. (1989) Use of thiol-terminal silanes and heterobifunctional crosslinkers for immobilization of antibodies on silica surfaces. *Anal. Biochem.* **178**, 408–413.

6

A Galactose-Specific Affinity Hollow Fiber Sensor Based on Fluorescence Resonance Energy Transfer

Ralph Ballerstadt and Jerome S. Schultz

1. Introduction

The quantitative detection of galactose or galactose-containing saccharides is of great importance in the food industry and in medical monitoring and treatment, as much as in ascertaining the basis of disease processes. For example, it was reported that disturbed galactose metabolism is connected with cataract formation in humans (1). Optical biosensors for galactose can have an impact on the medical practice. One approach was reported recently by Ballerstadt et al. (2), who described the potential use of a galactose-sensitive probe as a guidance system for the location and assessment of liver function in critically ill patients. The latter and several other prototypes of fiberoptic sensors (3,4) that have been developed for sugar sensing, e.g., glucose, are based on the use of lectins, a group of antibody-like, sugar-specific proteins. The use of enzymes for a fiberoptic biosensor has also been reported (5). In this chapter, we concentrate on the description of an optical method for the measurement of sugars based on the use of lectins as the affinity agent.

The reaction mechanism of this affinity assay consists of the competition between an analyte–analog (e.g., galactose-containing dextran) and the analyte (galactose) for the binding sites of the galactose-specific lectin *Ricinus communis* agglutinin. The biorecognition elements are enclosed inside a dialysis hollow fiber with a membrane porosity that allows nonhindered diffusion for galactose but prevents the outflow of the specific high-mol-wt reagents, i.e., the lectin and the analyte–analog.

The transduction of the competitive lectin–carbohydrate interaction into an electrical signal has been accomplished by employing fluorescence resonance

From: *Methods in Biotechnology, Vol. 7: Affinity Biosensors: Techniques and Protocols*
Edited by: K. R. Rogers and A. Mulchandani © Humana Press Inc., Totowa, NJ

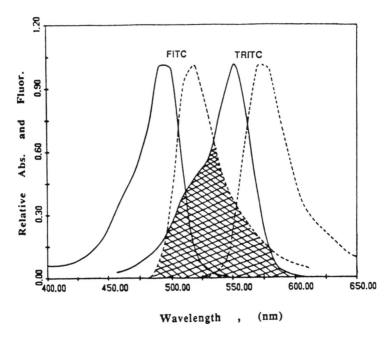

Fig. 1. Spectral overlap of the emission and absorption spectra of fluorescein and rhodamine.

energy transfer (FRET). This energy transfer (or quenching) occurs between a fluorescence donor dye (e.g., fluorescein) and a fluorescence acceptor dye (e.g., rhodamine) with an absorption spectrum of the fluorescence acceptor that overlaps to some extent with the emission spectrum of the fluorescence donor (**Fig. 1**). According to Förster's theory *(6)*, the efficiency of energy transfer, E, between the two moieties is given by

$$E = r^{-6}/(r^{-6} + R_o^{-6}), \text{ with } 0 \le E \le 1 \tag{1}$$

where r is the distance between the center of the donor and acceptor fluorochromes, and R_o is the distance at which the transfer efficiency is 50%. The distance between both fluorescein and rhodamine at which the fluorescence of fluorescein is quenched by 50% was found to be approx 56 Å *(7)*.

We used commercially available tetramethylrhodamine isothiocyanate (TRITC)-labeled dextran to prepare the macromolecular analog–analyte. To make this polymer a competitive ligand with galactose, the dextran was conjugated with lactose (a galactose-containing disaccharide). The reaction mechanism of the galactose-specific assay is as follows (*see* **Fig. 2**): Complexes of fluorescein isothiocyanate (FITC)-labeled RCAI (*Ricinus communis* agglutinin) and TRITC-labeled galactosyldextran are formed when galactose is absent. In

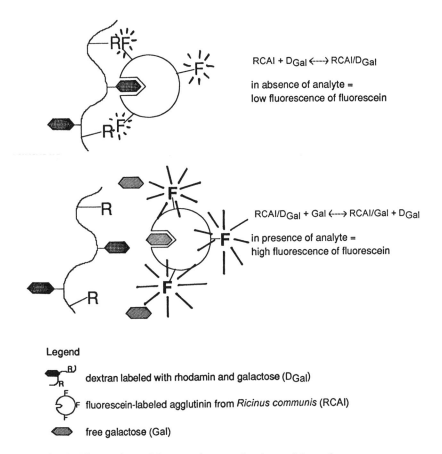

RCAl + D$_{Gal}$ $\longleftrightarrow$ RCAl/D$_{Gal}$

in absence of analyte =
low fluorescence of fluorescein

RCAl/D$_{Gal}$ + Gal $\longleftrightarrow$ RCAl/Gal + D$_{Gal}$

in presence of analyte =
high fluorescence of fluorescein

Legend

dextran labeled with rhodamin and galactose (D$_{Gal}$)

fluorescein-labeled agglutinin from *Ricinus communis* (RCAl)

free galactose (Gal)

Fig. 2. Illustration of the reaction mechanism of the galactose assay.

this structure the fluorescence donor and acceptor are in close proximity, resulting in quenching of the fluorescence. With increasing sugar concentration, the binding complex dissociates. This dissociation results in an increase in the distance between the fluorescein and rhodamine. At distances much greater than 50 Å, no quenching occurs and an increase in fluorescence is obtained.

The technique described in this chapter allows one to evaluate and to optimize the formulation of the sensor components without going through the arduous task of assembling the microscopic sensor units. Normally, the behavior of a sensor system is explored by making solutions containing the sensing reagents and the analyte at different concentrations and making measurements of aliquots with a fluorometer. These kinds of experiments give useful information about the relationship between assay composition and the analyte-induced changes in fluorescence signal. This method does not, however, truly

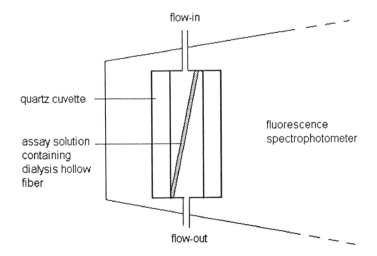

Fig. 3. Setup for monitoring the galactose-dependent fluorescence change of the assay solution enclosed inside a dialysis hollow fiber by means of a fluorescence spectrophotometer. The dialysis hollow-fiber segment was composed of regenerated cellulose and had a length of 1.3 cm, an inner diameter of 190 μm, and wall thickness of 20 μm. Both ends of the fiber were closed.

mimic sensor performance, i.e., one cannot continuously monitor analyte-induced fluorescence changes. Also, no insight is obtained on the important characteristics of time response and accuracy of repeated measurements.

To extract as much information as possible from the sensor assay within a short-time experiment, we devised a simple experimental setup consisting of a flowthrough cell into which a single short segment of a hollow fiber can be introduced (*see* **Fig. 3**). The entire unit is placed in a cuvet holder inside a desktop fluorometer. The flowthrough cell is perfused with the test solution containing the analyte (e.g., galactose), and the internal fluorescence change is monitored by the computer-controlled fluorometer. This configuration allows quick and easy removal and re-introduction of different hollow-fiber segments that are filled with the assay solution of interest. For instance, this is useful when the fiber is repeatedly transferred from the incubator into the flowthrough cell and back into the incubator as in tests of a sensor hollow fiber's long-term stability. Between these experiments other hollow-fiber sensors can be tested.

In fact, this method is generally applicable to other affinity assay systems (antibody–antigen) and bioelements (enzymes, microorganisms) in which the biorecognition reaction with a low-mol-wt analyte is associated with the emission of light and where the time-response and reversibility can be accurately tested without the effort of assembling fiberoptic devices.

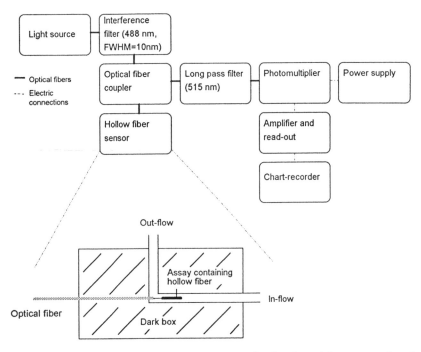

Fig. 4. Schematic illustration of the opticoelectronic circuit and the sensor chamber.

Eventually, when the optimal assay composition is found, it can be loaded into a small piece of hollow fiber that is attached at the tip of the optical fiber and the sensor function can be checked for remote sensing. The opticoelectronic circuit and the sensor chamber that have been used for this purpose are shown in **Fig. 4**. The galactose sensor is located inside a flowthrough cell, which is perfused with the test solution. The generated fluorescence is amplified by a photomultiplier tube (PAT) and monitored by a chart recorder.

2. Materials

All chemicals were from Sigma (St. Louis, MO), and used as received. Solutions of phosphate-buffered saline (PBS) had a pH of 7.2 and contained 8.3 mM of phosphate, pH 7.2, 0.9% NaCl, and 0.05% NaN$_3$ (stabilizer).

2.1. Labeling of Ricinus communis *Agglutinin (RCAI) with FITC*

1. *Ricinus communis* agglutinin, dissolved in 0.005 M Na$_2$HPO$_4$, 0.2 M NaCl, pH 7.2, 0.1% NaN$_3$, Lot#115h4033, protein concentration 8.8 mg/mL, RCAI is a toxic protein and can cause allergic reactions; wear gloves.
2. 1 mg Fluorescein isothiocyanate (FITC) isomer II.
3. 1 M Sodium bicarbonate (NaHCO$_3$).

4. Dialysis tubing (Spectra/Por MWCO 25 000, Spectrum, Houston, TX).
5. 20 mM D-(+)-Galactose.

2.2. Conjugation of RITC-Dextran with Lactose

1. 35 mg Tetramethylrhodamine isothiocyante (TRITC)-dextran (150 kDa, 0.008 mol dye/mol glucose).
2. 200 mg Lactose.
3. 5% (w/v) Glycine.
4. Divinyl sulfone ($C_4H_6O_2S$, DVS), DVS is highly toxic and should be handled with care. Wear gloves and work under a hood.
5. Buffer for galactose-grafting of TRITC-dextran: 1 M Na_2CO_3.
6. Freeze-drying machine.
7. 1 M Hydrochloride (HCl).

2.3. Preparation of a Sensor Fiber Segment and Its Encasement Inside the Flowthrough Cell

1. Assay solution: 0.6 mg/mL FITC-labeled RCAI and 0.6 mg/mL TRITC-galactosyldextran dissolved in PBS.
2. Dialysis hollow fibers (Kunstseidewerk, Pirna, Germany) composed of regenerated cellulose, id 190 μm, wall thickness 20 μm, cut-off mol wt 10,000.
3. Adhesive based on cyanacrylate (Bel-Arts Products, Pequannock, NJ). Since cyanacrylate can cause eye injury, care must be taken.

2.4. Measurement of the Galactose-Dependent Fluorescence Change

1. Assay solution: 0.6 mg/mL FITC-labeled RCAI and 0.6 mg/mL TRITC-galactosyldextran dissolved in PBS
2. 0–2 mM Galactose dissolved in PBS.
3. Computer-controlled desktop fluorescence spectrophotometer LS50B (Perkin-Elmer, Beaconsville, UK).
4. Flowthrough cell (LC flow cell accessory for LS50B, part-number: L225-0138).

2.5. Preparation of the Fiber Optic Sensor and Components of the Opticoelectronic Circuit

1. Hollow fibers, adhesive, and assay solution as above (see **Subheading 2.3.**).
2. Light source fiberlite (high intensities illuminator series 180, Dolan-Jenner Industries, Lawrence, MA).
3. Interference filter (1.0-in diameter, 488 nm, full width at half maximum (FWHM) = 10 nm, Newport, Irvine, CA).
4. Multimode single optical fibers (Siecor, core 100 μm, cladding 140 nm).
5. Fiber coupler (percentage of light passing through: light source to sensor 19%, sensor to PMT 75%, light source to PMT <0.01%, Canstar, Scarborough, Ontario, Canada).
6. 515 nm Long-pass filter.
7. Head-on photomultiplier (Oriel, Stratford, CT).

8. Power supply for PMT (Type #7070, Oriel).
9. Current preamplifier (70710, Oriel).
10. Readout (70710, Oriel).
11. Linear chart recorder (VWR, USA).

3. Methods

3.1. Labeling of Ricinus communis Agglutinin (RCAI) with FITC

1. Dilute 0.5 mL of the RCAI preparation with 0.2 mL of 1 M NaHCO$_3$ in a 1.5-mL polypropylene tube.
2. Add solid galactose (final concentration 20 mM).
3. Add 1 mg of FITC to the solution. Allow the reaction to proceed for 1–2 h at room temperature. Store the reaction tube at 4°C in a dark room overnight.
4. Separate the FITC-RCAI conjugate from unbound FITC by equilibrium dialysis against 2 × 4 L of PBS overnight.
5. Store the dialysate at 4°C in the dark.

3.2. Conjugation of RITC-Dextran With Lactose

1. Dissolve 35 mg of TRITC-dextran in 2 mL distilled water, and add 2 mL of 1 M Na$_2$CO$_3$.
2. During intensive stirring, add 30 µL (activation step) of DVS to the solution and allow the reaction to proceed for 1 h.
3. Add a saturating amount of lactose (200 mg) to the mixture and allow the reaction to proceed for 1 h (coupling step).
4. Add glycine (5% w/v) to the stirred solution and adjust the pH of the solution to 9.0–10.0 by 1 M HCl (blocking step, 1 h).
5. Transfer the dextran solution to a dialysis bag and dialyze against 4 L of distilled water containing 0.5% (w/v) glycine and 0.1% (w/v) NaN$_3$ for 4–6 h, and thereafter against PBS overnight.
6. Centrifuge the dialysate briefly at 10,000g for 10 min and pour the supernatant into a dust-free glass vessel.
7. Store the dextran solution at –80°C for 1 h and lyophilize the frozen preparation by means of a freeze-drying machine.
8. Store the freeze-dried dextran at 4°C in the dark.

3.3. Preparation of a Sensor Fiber Segment and Its Encasement Inside the Flowthrough Cell

1. Cut a dry hollow fiber in small segments of 2–3 cm in length.
2. Aspirate the assay solution composed of FITC-RCAI and TRITC-galacto-syldextran (for concentrations, *see* **Subheading 2.3.**) into the hollow fiber by putting the fiber inside the assay solution (0.2 mL).
3. Close the upper fiber end with cyanacrylate and compress with tweezers.
4. Remove the hollow-fiber segment from the tube. Quickly cut it to a length of 1.5 cm and immediately close with cyanacrylate. Store the fiber segments in PBS.

Carry out the closing of the assay solution-loaded hollow fiber as quickly as possible to avoid loss of solution resulting from evaporation. It is recommended that this step be performed using a stereomicroscope.

5. Carefully introduce the hollow fiber into a quartz cuvet with tweezers and push it down with the inflow cover of the flowthrough cell unit. This results in a slight bending of the hollow fiber that maintains it in a fixed position.

3.4. Measurement of the Galactose-Dependent Fluorescence Change

1. Spike the PBS-solution with different galactose concentrations ranging from 0 to 2 mM.
2. Start the measurements by perfusing PBS without galactose through the measuring cell to get a baseline. Continue the measurements by changing the galactose concentration of the perfusing solution. Use a pump or simple gravity to flow the test solution through the cell. When the effect of galactose on the internal fluorescence change is investigated, its nonspecific contributions to the galactose-generated fluorescence signal (e.g., refractive index, or trace fluorescence) have to be checked by measuring the internal fluorescence of an assay solution without lectin.

3.5. Preparation of the Fiberoptic Sensor

1. Strip off the fiber jacket at the end of the optical fiber (2 cm). Remove the cladding by soaking in acetone (2 min) and pull it off with tweezers.
2. To get an optimal fluorescence signal, a mirrorlike surface of the fiber end is essential. Bend the bare fiber slightly. While applying tension, score the fiber with a sharp razor blade, causing the fiber to break. Examine the quality of the fiber end under a stereomicroscope.
3. Carefully push a small piece of the hollow fiber segment (2–3 cm) on the prepared bare fiber end (2–4 mm), and cut the protruding end of the hollow fiber to a length of ca. 5 mm using scissors.
4. Aspirate the assay solution (for composition and concentrations, *see* **Subheading 2.4.**) into the hollow fiber by dipping the open end of the fiber into a drop of this solution. Rapidly close both ends with cyanacrylate.
5. Position the prepared sensor fiber into the flowthrough chamber. Cover the chamber with a light-proof cap. Connect the other end of the optical fiber with the fiber coupler.
6. Start the measurements by pumping PBS solution through the flowthrough cell spiked with different galactose concentrations ranging from 0–2 mM, and registrate the change in fluorescence by using a chart recorder.

4. Notes

1. Labeling of *Ricinus communis* agglutinin (RCAI) with FITC: The addition of a saturating amount of galactose to the solution prevents the lectin from losing binding activity by protecting the lectin's sugar binding site for covalent binding of fluorescein.

   To purify the FITC-RCAI conjugate, separation techniques (e.g., size-exclusion chromatography or ultrafiltration) other than equilibrium dialysis, as shown in **Subheading 3.1., step 4,** may be used. It should be noted that the separation by dialysis is not quite complete. Incubating the hollow-fiber segments that contain the FITC-RCAI or TRITC-galactosyldextran solution in PBS removes unbound FITC residues.

2. Grafting of TRITC-dextran with lactose: Divinyl sulfone is a bifunctional crosslinking agent and reacts specifically with —OH, —NH, and —SH groups. It was first used by Porath and coworkers *(8)*, who grafted agar with sugars. Their method was modified slightly by Ehwald et al. *(9)* to couple covalently soluble dextran with different sugars and insulin. The latter protocol was used as described here. TRITC-dextran was employed as an analyte–analog because of its good solubility, its commercial availability in a variety of different sizes, and its well-defined labeling parameters. It was found that the efficiency of fluorescence quenching of TRITC-dextran after affinity binding by FITC-RCAI depends heavily on the TRITC–glucose mass ratio of the used dextran, which should be as high as possible. However, the molecular weight has no influence on fluorescence quenching.

   The concentration of TRITC used here was found to give the best results with regard to the amount of fluorescence change of the sensor assay after galactose reaction. Higher dextran concentrations are not recommended because of the possibility of crosslinking of the TRITC-dextran. At lower dextran concentrations, the time for activation has to be prolonged. Intensive stirring is essential for good mass convection as well as crosslinking prevention.

3. Preparation of sensor fiber and its encasement inside the flowthrough cell of the fluorescence spectrophotometer: A variety of hollow fibers, differing in their size, chemical composition, strength, and so forth, are available on the market for in vivo use. For the galactose sensor, we preferred to use hollow fibers made of regenerated cellulose, because they have a well-known history in in vivo applications, e.g., kidney dialysis.

Acknowledgment

This work was supported by a grant from the German Exchange Service (DAAD).

References

1. Birlouz-Aragon, I., Ravelontseheno, L., Villate-Cathelineau, G., Cathelineau, G., and Abitbol, G. (1993) Disturbed galactose metabolism in elderly and diabetic humans is associated with cataract formation. *J. Nutr.* **123,** 1370–1376.
2. Ballerstadt, R., Dahn, M., Schultz, J. S., and Lange, P. (1996) A homogeneous fluorescence assay system for galactose monitoring. 3rd European Conference on Optical and Biosensors, Zurich, Switzerland.
3. Schultz, J. S., Mansouri, S., and Goldstein, I. J. (1982) Affinity sensor: a new technique for developing implantable sensors for glucose and other metabolites. *Diabetes Care* **5,** 245–253.
4. Meadows, D. L. and Schultz, J. S. (1993) Design, manufacture and characterization of an optical fiber glucose affinity sensor based on a homogeneous fluorescence energy transfer assay system. *Anal. Chim. Acta* **280,** 21–30.

5. Rosenzweig, Z. and Kopelman, R. (1996) Analytical properties and sensor size effects of a micrometer-sized optical fiber glucose biosensor. *Anal. Chem.* **68,** 1408–1413.
6. Förster, T. (1960) Transfer mechanisms of electronic excitation energy. *Rad. Res. Supp.* **2,** 326.
7. Johnson, D. A., Voet, J. G., and Tayler, P. (1984) Fluoresence energy transfer between cobra α-toxin molecules bound to acetylcholine receptor. *J. Biol. Chem.* **259,** 5717.
8. Porath, J., Laas, T., and Janson, J.-C. (1975) Agar derivatives for chromatography, electrophoresis and gel-bound enzymes. II. Rigid agarose gels cross-linked with divinyl sulphone (DVS). *J. Chromatogr.* **103,** 49–62.
9. Ehwald, R., Ballerstadt, R., and Dautzenberg, H. (1996) Viscosimetric affinity assay. *Anal. Biochem.* **234,** 1–8.

7

Fiberoptic Immunosensors with Continuous Analyte Response

J. Rex Astles, W. Greg Miller, C. Michael Hanbury, and F. Philip Anderson

1. Introduction

1.1. Background

Fiberoptic fluorescence signal transmission has several advantages for immunosensor design: physical flexibility for remote sensing, no risk of electrical interference, high signal-to-noise ratio with little attenuation over distance, and the capacity to both measure several analytes with fluorescence from a single fiber and bundle fibers without significant crosstalk. For many immunosensors these possibilities have offered little advantage because they have been designed for single use, or they have required regeneration that usually can not be accomplished *in situ*. The sensor described here, developed by Anderson and Miller *(1)*, is self-contained and completely reversible because the antibody has a sufficiently fast effective dissociation rate constant (k_{dis}). The sensor can be used for hours to days depending on the application. It has been calibrated and used in blood to measure therapeutic concentrations of free phenytoin (PHT) *(2)*, and the design can be modified for use with other haptens, such as theophylline (THEO) *(3)*. Here we present general instructions for preparation of reversible fiberoptic immunosensors, as well as specific details for construction of sensors to phenytoin and theophylline.

1.2. Basis of the Immunosensor

The indicator system is membrane-encapsulated at the end of an optical fiber and depends on changes in fluorescence energy transfer between the donor, B-phycoerythrin (BPE), and the acceptor, Texas Red (TR). The inten-

From: *Methods in Biotechnology, Vol. 7: Affinity Biosensors: Techniques and Protocols*
Edited by: K. R. Rogers and A. Mulchandani © Humana Press Inc., Totowa, NJ

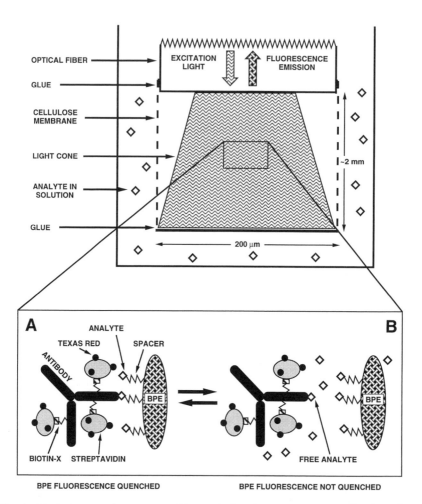

Fig. 1. **(Top)** A sensor is constructed by enclosing the reagents, TR-antibody and BPE-antigen, inside a short segment of cellulose dialysis membrane at the end of a fiberoptic. **(Bottom)** reagent response to analyte, ◇. In the absence of exogenous analyte **(A)**, the BPE-PHT is maximally quenched by Texas Red-antibody. In the presence of exogenous analyte **(B)**, less Texas Red-antibody is bound to analyte and the BPE fluorescence signal increases.

sity of fluorescence from BPE labeled with phenytoin (BPE-PHT) or other analyte is decreased when the antibody, labeled with TR, binds to phenytoin. **Figure 1A** shows the condition when antibody–hapten binding places the BPE and TR in close proximate geometry for efficient energy transfer and thus quench of BPE fluorescence. In the absence of analyte phenytoin, antibody molecules are fully packed around BPE and fluorescence quenching is maximal. When free phenytoin is present it competes with the BPE-PHT for TR-antibody. A portion of the TR-antibody binds to analyte phenytoin, making it unavailable to quench BPE, increasing the fluorescence signal (**Fig. 1B**). At equilibrium, the concentration of free phenytoin in the test solution produces a characteristic fraction of TR-antibody bound to BPE-PHT, with a unique and reproducible fluorescence intensity.

In this design the main determinants of response time are the effective k_{dis}, the reagent solution viscosity, and membrane interactions. Simulations indicate that the association rate constant (k_{ass}) has little impact on the time to reach equilibrium after a change in analyte concentration, whereas a large k_{dis} is essential for a rapid response *(4)*. Changing k_{ass} by three orders of magnitude has little effect on the simulated response time. However, if the k_{ass} is held constant at 4×10^6 L/mol/s, increasing the k_{dis} one order of magnitude, for example from 4×10^{-4} to 4×10^{-3} s^{-1}, results in a 10-fold reduction in response time (162 to 16 min). Because k_{dis} rates may vary by as many as eight orders of magnitude *(5)*, antibodies differ in suitability for continuous measurement systems and should be chosen judiciously to achieve an appropriate sensor response time.

The phenytoin sensor described here uses antibodies with a k_{dis} of 4×10^{-3} s^{-1}, and intact sensors have response times of 15 m in aqueous solution *(2)*. Reagent viscosity can impact response time because the free analyte must diffuse into and throughout the reagent chamber to affect a change in fluorescence. For example, when dextran 70 kDa is included in the sensor for plasma or blood to balance oncotic pressure in the specimen, the response time is 1 h *(2)*. Because of viscosity effects, and other unpredictable interactions between reagents and the encapsulation membrane *(3)*, sensor performance may differ substantially from theoretical. The best approach for sensor design may be to select antibodies with a relatively large k_{dis} and satisfactory specificity, then test intact sensors to determine if the response time is appropriate.

Encapsulation of the reagent system with a dialysis membrane ensures that only nonprotein–bound analyte is measured and also isolates the system from high-mol-wt interfering substances. The membrane defines the volume in which reagents are distributed and this volume can change depending on the oncotic pressure gradient across it. Therefore, care should be used to match the oncotic pressure between the reagents and the test solution (*see* **Subheading 3.1.4.**, oncotic pressure equilibration).

2. Materials

2.1. Instrumentation

The optical components of the immunosensor are shown in **Fig. 2.**

1. Excitation source: a 100-mW air-cooled argon ion laser (model 5500A from Ion Laser Technology, Salt Lake City, UT) tuned to 514.5 nm and attenuated to 0.1% with a neutral-density filter.

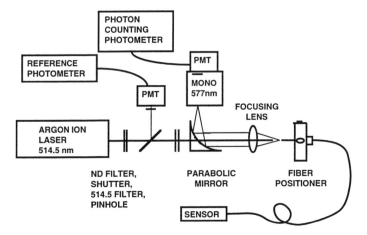

Fig. 2. The fiberoptic fluorometer.

2. Reference channel beam splitter (conventional glass microscope slide at 45° to the incident beam) and photometer.
3. Off axis parabolic mirror (#02POA015, Melles Griot, Irvine, CA) of 66-mm focal length with a 2-mm hole drilled on the optical axis.
4. Focusing lens, fused silica with a 25-mm focal length (#01LQB028, Melles Griot).
5. Fiber positioner (#FP-2, Newport, Fountain Valley, CA).
6. Emission monochrometer set at 577 nm.
7. Optical fiber 100/110 μm core/cladding fiber (Superguide G #B2–0007-20, Fiberguide Industries, Stirling, NJ).
8. The remaining optical components (breadboard, photomultiplier tubes, photometers, lens mounts, shutter, and filters) are standard.

2.2. Reagents

Sensor reagents consist of (1) BPE labeled with the analyte of interest via a spacer, (2) the appropriate antibody labeled either directly with fluorescence energy acceptor or indirectly with an acceptor attached to streptavidin—in the latter case the antibody is first biotinylated, and (3) as necessary, the addition of an agent (dextran) to balance the oncotic pressure if the sensor is to be used in a solution with appreciable protein content, such as blood.

2.2.1. Sensor for Phenytoin (BBE-Phenytoin)

The following solvents are needed: *N, N*-dimethylformamide (DMF); ethyl acetate; 1,4 dioxane; HCl, deionized water (DW). Dry DMF with molecular sieves (20,860-4, Aldrich, Milwaukee, WI).

2.2.1.1. SYNTHESIS OF PHENYTOIN-VALERATE (PHT-VAL)

1. Sodium phenytoin, 2.74 g (Sigma, St. Louis, MO).
2. Ethyl 5-bromovalerate, 2.09 g (Aldrich).
3. Anhydrous calcium sulfate desiccant.

2.2.1.2. PHENYTOIN-VALERATE-NHS (PHT-VAL-NHS) (*SEE* **NOTE 2**)

1. PHT-Val, 176 mg (0.5 mmol) prepared according to **Subheading 3.1.1.2.**
2. Dicyclohexyl-carbodiimide, 113 mg (0.55 mmol) (DCC, Aldrich).
3. *N*-hydroxysuccinimide, 69 mg (0.6 mmol) (NHS, Aldrich).

2.2.1.3. PHENYTOIN-VALERATE-BPE (BPE-PHT)

1. BPE (Molecular Probes, Eugene, OR), 0.29 g/L in phosphate-buffered saline (PBS: 10 mM sodium phosphate and 150 mM NaCl, pH 7.4), prepared according to **Subheading 3.1.1.4.**
2. PHT-Val-NHS in dioxane, 1 g/L, prepared according to **Subheading 3.1.1.3.**

2.2.2. Purified Antiphenytoin Antibody

Refer to comments in **Subheading 3.1.2.** to choose a method for antibody purification.

2.2.2.1. SALTING OUT

1. Anti-PHT mouse monoclonal IgG in ascites fluid (clone 157/11; Chemicon International, El Segundo, CA).
2. Saturated ammonium sulfate.
3. Borate-buffered saline (BBS, 10 mM boric acid and 150 mM NaCl, pH 9.0).

2.2.2.2. ALTERNATIVE PURIFICATION: PROTEIN A CHROMATOGRAPHY

1. Anti-PHT monoclonal mouse IgG in ascites fluid (clone 157/11; Chemicon).
2. Antibody purification kit (Pierce, Rockford, IL).
3. Dialysis tubing, 10–20 kDa cutoff, and excess PBS.
4. Materials for protein concentration, for example by centrifugal filtration.

2.2.2.3. ANTIBODY BIOTINYLATION

1. Purified antibody in BBS, approx 1–2 mg/mL.
2. Biotin-aminocaproyl-*N*-hydroxysuccinimide (biotin-X-NHS, Aldrich) in anhydrous DMF.
3. Dialysis tubing, 10–20 k cutoff, and excess PBS.

2.2.3. Texas Red–Streptavidin Preparation

Texas Red–streptavidin can be obtained preconjugated (Molecular Probes) or prepared as described in **Subheading 3.1.3.**, and **Note 3c**.

1. Streptavidin (Molecular Probes), 5 mM in 50 mM borate buffer, pH 9.1.
2. Texas Red sulfonyl chloride, 0.5 mg in 50 μL anhydrous DMF.
3. Sephadex G-25 in 1 × 15-cm column equilibrated with PBS.
4. Dialysis tubing, 10–20 kDa cutoff, and excess PBS.
5. A pyroreactor (Antek Instruments, Houston, TX) nitrogen detector (or *see* **Subheading 3.1.3.**).

2.2.4. Oncotic Pressure Equilibration (Measurements in Blood)

1. Dextran 70 kDa (Sigma) in 150 mM PBS: 0, 20, 40, 60, 80, and 100 g/L.
2. Plastic, capped centrifuge tubes.
3. Pooled plasma, 10 mL.
4. Dialysis membrane, 1-cm diameter, molecular weight cutoff matched to sensor.
5. Spectrophotometer set at 280 nm.

2.2.5. Final Reagent Assembly

Reagent concentrations must be optimized (*see* **Subheading 3.3.**).

1. Small glass centrifuge tube, magnetic "flea," and magnetic mixer.
2. PBS buffer.
3. BPE-PHT, 10 nM.
4. Biotinylated antibody, 1 μM.
5. Texas Red-streptavidin, 10 μM.
6. Dextran 70 kDa stock solution (if necessary).

2.2.6. Sensor for Theophylline (BPE-Theophylline)

Materials for preparation of the theophylline sensor are exactly as described for the phenytoin sensor except for the preparation of BPE-theophylline (BPE-THEO). The following solvents are needed: DMF; deionized water. Dry DMF with molecular sieves (Aldrich).

2.2.6.1. THEOPHYLLINE–NITROPHENOL ACTIVATED ESTER (THEO-O-PNP)

1. 8-3-Carboxypropoyl-1,3-dimethyl-xanthine, 30 mg (C4041, Sigma).
2. *p*-Nitrophenol, 47.3 mg (PNP, Aldrich).
3. Dicyclohexyl-carbodiimide, 70.1 mg (Aldrich)
4. Triethylamine, 5.7 mg.

2.2.6.2. BPE-THEOPHYLLINE (BPE-THEO)

1. BPE, 2.0 mg/mL in PBS, prepared as described in **Subheading 3.1.1.4.**
2. THEO-O-PNP, 200 μL in DMF at >4000-fold molar excess relative to BPE.
3. Dialysis tubing, 10–20 kDa cutoff, and excess PBS.

2.2.7. Antitheophylline Antibody

1. See **Subheadings 3.1.** and **3.1.7.**

2.2.8. Final Reagent Assembly

Concentrations must be optimized (*see* **Subheading 3.3.**).

1. Small glass centrifuge tube, magnetic flea, and magnetic mixer.
2. PBS buffer.
3. BPE-antigen, 20 n*M*.
4. Biotinylated antibody, 60 n*M*.
5. Texas Red–streptavidin, 160 n*M*.
6. Dextran 70 kDa stock solution (measurements in blood).

2.3. Sensor Construction and Use

2.3.1. Sensor Assembly

1. FK11 Fiber cleaver (York Technology, Princeton, NJ).
2. Sigmacote™ (Sigma).
3. Hollow fiber dialysis membrane (10-kDa molecular weight cutoff) from a bioreactor (Cell-Pharm, #BR110, CD Medical, Miami Lakes, FL).
4. Tygon™ tubing with inner diameter slightly larger than the dialysis tubing.
5. Rubber cement.
6. Tuberculin syringe, 1-mL with small-gage needle.
7. Cyanoacrylate glue.
8. Surgical scissors.

2.3.2. Calibration

2.3.2.1. SECONDARY STANDARDS

1. Plasma pool spiked with various concentrations PHT (or other antigen).
2. Centrifugal separation device (Micropartition System with a YMT 14-mm membrane, Amicon, Danvers, MA).
3. See **ref. 3** for materials necessary for assignment of secondary standard concentrations.

2.3.2.2. PRIMARY STANDARDS

1. Dextran 70 kDa or other oncotic agent (BSA, and so forth). *See* **Subheading 3.5.1.1.** to test compatibility of oncotic agent with primary standards.

3. Methods

3.1. Sensor Reagent Preparation

3.1.1. Sensor for Phenytoin (BPE-Phenytoin)

Preparation of BPE-PHT is done in stages:

1. A spacer is attached to phenytoin creating phenytoin-valeric-ethyl ester;
2. The ester is hydrolyzed to the corresponding acid, phenytoin-valeric acid (PHT-Val);
3. The *N*-hydroxysuccinimide derivative (PHT-Val-NHS) is prepared; and
4. BPE is reacted with PHT-Val-NHS.

3.1.1.1. Phenytoin-Valeric-Ethyl Ester

1. Dissolve sodium phenytoin (2.74 g) and ethyl 5-bromovalerate (2.09 g) in anhydrous DMF (100 mL). Sonicate in a covered ultrasonic cleaner at 65°C for 4 h.
2. Concentrate DMF solution in a high vacuum evaporator to about 2 mL.
3. Add DW (50 mL) to isolate the ester as a yellow oil and stir at room temperature for 12 h to convert to a white powder.
4. Wash the product over a sintered glass funnel with DW (500 mL) and dry over anhydrous calcium sulfate.

3.1.1.2. Phenytoin-Valerate (PHT-Val)

1. Reflux the above ester for 2 h in 10% aqueous 1,4 dioxane with 0.5 M HCl (50 mL).
2. Add cold DW (500 mL) to precipitate the acid and wash with cold DW as above.
3. To remove water, reflux the acid in ethyl acetate (20 mL) in a 100-mL Erlenmeyer flask over a hot plate for 10 min. Recrystallize at 4°C overnight. The melting point is 155–156°C *(6)*.

3.1.1.3. Phenytoin-valerate-NHS (PHT-Val-NHS) (*see* Note 2)

1. Dissolve PHT-Val (176 mg, 0.5 mmol), DCC (113 mg, 0.55 mmol), and NHS (69 mg, 0.6 mmol) in dioxane (20 mL) and stir at room temperature for 4 h.
2. Calculate the concentration of PHT-Val-NHS assuming complete conversion of the 0.5 mmol PHT-Val to PHT-Val-NHS.
3. Adjust the concentration of PHT-Val-NHS to 1 g/L with dioxane.

3.1.1.4. Phenytoin-Valerate-BPE (BPE-PHT)

1. Centrifuge to recover BPE suspension from the ammonium sulfate used for storage. Discard the supernatant and dissolve the pellet in an equal volume of phosphate buffer (PBS: 10 mM/L sodium phosphate and 150 mM NaCl, pH 7.4).
2. Dialyze against PBS (1000-fold excess), three times at 4°C for 12 h.
3. Measure the concentration of BPE using the molar absorptivity of 2.41×10^6 AL/mol/cm at 540 nm *(7)* and adjust the final concentration of the BPE to 0.29 g/L in PBS.
4. Add 33 µL PHT-Val-NHS in dioxane (1 g/L) to 625 µL of BPE in PBS (0.29 g/L) and stir gently at 4°C overnight.
5. Dialyze the product, BPE-Val-PHT, against PBS (1000-fold excess) at 4°C for 12 h times 3, and use the molar absorptivity of native BPE, 2.41×10^6 AL/mol/cm at 540 nm, to measure the concentration. Subsequent references to BPE-PHT conjugate concentration refer to the concentration of BPE; the number of PHT moieties per BPE cannot easily be quantified.

3.1.2. Antiphenytoin Antibody

3.1.2.1. Antibody Purification

There are two alternatives for antibody purification: (1) salting out with ammonium sulfate is fast, but may leave some impurities, including albumin and possibly proteases, or (2) affinity chromatography with protein A, which takes longer and may have a lower yield (70%), but produces a purer preparation.

3.1.2.1.1. Method 1: Ammonium Sulfate Precipitation

1. While stirring 1.0 mL ascites fluid in a small test tube at room temperature, very slowly add 0.8 mL saturated ammonium sulfate. Stir for 30 min at room temperature.
2. Centrifuge the precipitated antibody in a small test tube at 2300g for 20 min. Discard the supernatant and add 1.0 mL to dissolve the precipitate. Dialyze against BBS (1000-fold excess) at 4°C for 12 h times 3.
3. Determine IgG concentration from the molar absorptivity 2.1 × 10^5 AL/mol/cm at 280 nm *(8)*.

3.1.2.1.2. Method 2: Protein A Chromatography

1. Load ascites fluid (0.5 mL) on a 1- × 3-cm protein A column pre-equilibrated with "binding buffer," pH 8.3, that is included in the antibody purification kit.
2. Wash the column with 5 mL binding buffer, and collect 1-mL fractions, monitoring the absorbance at 280 nm. Wash with an additional 5 mL of binding buffer after the absorbance returns to baseline (absorbance < 0.005).
3. Recover the IgG in 1-mL fractions using "elution buffer," pH 2.8. Measure absorbance in the fractions at 280 nm, and use the molar absorptivity of IgG, 2.1 × 10^5 AL/mol/cm, to estimate recovery. About 2.3 mg IgG should be recovered in 5–7 mL (0.40 g/L).
4. Concentrate the antibody to about 1.5–2.0 mg/mL using a standard method for protein concentration. Determine antibody concentrations from the molar absorptivity, as above, after final dialysis against pH 9.0 BBS in preparation for biotinylation.

3.1.2.2. ANTIBODY BIOTINYLATION

For this clone (Chemicon 157/11), biotin–streptavidin coupling is necessary to avoid antibody denaturation by direct attachment of Texas Red. The tendency to precipitate with direct labeling may depend on the particular antibody.

For biotinylation, use biotin-X-NHS to antibody ratio of 50:1 if performed at pH 9.0. *See* **Note 3b** and **Subheading 3.3.1.** to optimize conditions.

1. Prepare biotin-X-NHS in anhydrous DMF to be 50-fold more concentrated (molar) than antibody.
2. Add biotin-X-NHS to the stirred antibody solution (5% v/v) at room temperature and allow to react for 2 h. For example, 50 μL of 4.3 mg/mL biotin-X-NHS (450 nmol) is added to 1 mL antibody that is 1.4 mg/mL (9.3 nmol).
3. Dialyze against BBS (1000-fold excess) three times at 4°C for 12 h, and measure concentration as above.

3.1.3. Texas Red-Streptavidin Preparation

Streptavidin has a lower pI than avidin, is not as positively charged at physiologic pH, and is therefore less apt to aggregate and precipitate with the other reagent proteins. Despite a higher cost, using streptavidin may be preferable when reagent stability is a problem. Although preparation of Texas Red-streptavidin from the sulfonyl chloride is straightforward, an excess of the labeling reagent can cause precipitation (*see* **Note 3c**), making commercially obtainable conjugate an attractive alternative.

When streptavidin is labeled with Texas Red the labeling efficiency should be assessed. When labeling conditions are optimal, the ratio of Texas Red to streptavidin is about 3–3.5 (*see* **Note 3c** for alternative ways to measure the labeling efficiency).

1. Dissolve streptavidin (5 mg) in 1 mL of 50 mmol/L borate buffer, pH 9.1, and cool on ice. Dissolve Texas Red sulfonyl chloride (0.5 mg) in 50 µL anhydrous DMF (prepare fresh).
2. Immediately add the Texas Red reagent to the streptavidin solution and stir for 1 h at 4°C.
3. Purify the conjugated streptavidin on a 1- × 15-cm column of Sephadex G-25, eluting with PBS. Separate the blue conjugated protein from the unreacted Texas Red and dialyze against PBS three times (1000-fold excess).
4. Measure the concentration of the labeled streptavidin as total nitrogen, assuming 6.25 g protein per gram of nitrogen *(10)*. A pyroreactor calibrated with ammonium sulfate standards can be used to measure total nitrogen. This method combusts the diluted protein solution at 1110°C with oxygen to produce nitric oxide, which reacts with ozone to produce chemiluminescence proportional to nitrogen content. Kjeldahl or other methods can also be used.

3.1.4. Oncotic Pressure Equilibration (Measurements in Blood)

Determine the appropriate concentration of dextran empirically by estimating at what dextran concentration there would be no net migration of water between the reagents and plasma across the sensor dialysis membrane. The molecular weight cutoff and composition of the dialysis membrane should match the membrane used in the sensor.

1. Prepare several concentrations of dextran 70 kDa in 150 mmol/L PBS: 0, 20, 40, 60, 80, and 100 g/L.
2. Dispense 50 mL of each dextran solution into a plastic, capped centrifuge tube.
3. Prepare a 10-mL pool of plasma and determine the absorbance of a 1:100 dilution at 280 nm. Aliquot 1 mL of plasma pool into short segments of 1-cm diameter dialysis membrane.
4. Place one segment of plasma pool into each centrifuge tube and determine the absorbance of 1:100 dilutions of the dialysates after a 12 h incubation at room temperature (or other temperature appropriate for the intended use).
5. Plot the absorbance at 280 nm (*y*-axis) vs dextran concentration (*x*-axis) and fit the curve with a two-parameter polynomial expansion or other suitable method.

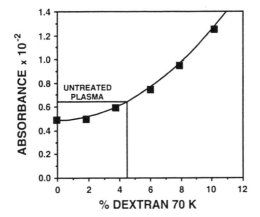

Fig. 3. Empirical estimation of the appropriate dextran 70 kDa concentration to balance the oncotic pressure. The absorbance at 280 nm was used to measure protein content of aliquots of plasma in dialysis tubing after equilibration. In the presence of 4.5% dextran 70 kDa the oncotic pressure is balanced. There is no net movement of water so absorbance is the same as for plasma without manipulation (0.65).

6. From the graph estimate the dextran concentration where the initial absorbance of the unmanipulated pool and the absorbance of the dialysate are equal (4.5% in **Fig. 3**).
7. Prepare a stock solution of dextran 70 kDa that is appropriately concentrated so that the final concentration in the sensor reagents will be correct. A 1:1 final dilution of a twofold (9%) concentrated stock works well.

3.1.5. Final Reagent Assembly

Mix the reagent components together immediately before sensor preparation. A small test tube with a magnetic flea stirrer is convenient to mix the reagents as they are added sequentially. At least two components are needed for use in protein-free buffer: BPE-antigen and TR-antibody. If PBS or other buffer is needed to dilute the reagents, add it first, then the BPE-antigen, and then the antibody-acceptor. If Texas Red–streptavidin is used to label biotinylated antibody, add it third. *See* **Note 4** if using TR–streptavidin instead of directly labeled antibody. Lastly, for sensor work in blood, add the dextran 70 kDa stock solution. Although there may be some precipitation, the reagents will function in a sensor. Do not centrifuge the reagent suspension.

Generally, 500 µL reagent total volume is sufficient to assemble a working sensor even if initial attempts fail because of leakage. Final concentrations of reagents are 10 nM BPE-PHT, 1 µM antibody, and 10 µM Texas Red–streptavidin in PBS with 4.5% dextran 70 kDa. However, optimize the reagent concentrations according to **Subheading 3.3.** and **Notes 3a–c**. The specific volumes to add will depend on the stock concentrations.

3.1.6. Sensor for Theophylline (BPE-Theophylline)

The analytical system and preparation of the reagents for the theophylline sensor are exactly as described for the phenytoin sensor except for the preparation of BPE-theophylline (BPE-THEO). BPE-THEO is prepared in two steps. The theophylline derivative, 8-3-carboxypropoyl-1,3-dimethyl-xanthine, has a 5-carbon side chain spacer arm with a terminal carboxylic acid. It is reacted with p-nitrophenol (PNP) in the presence of dicyclohexyl-carbodiimide and triethylamine to form the activated ester, THEO-O-PNP, which is then reacted with BPE.

3.1.6.1. THEOPHYLLINE–p-NITROPHENOL ACTIVATED ESTER (THEO-O-PNP)

1. In anhydrous DMF (1 mL), dissolve p-nitrophenol (47.3 mg, 340 μmol), dicyclohexyl-carbodiimide (70.1 mg, 340 μmol), triethylamine (5.7 mg, 56 μmol), and 8-3-carboxypropyl-1,3-dimethyl-xanthine (30 mg, 113 μmol). Stir at 4°C for 4 h, then overnight at room temperature. Completion of the reaction can be followed with high-pressure liquid chromatography (HPLC) and the yield is typically about 60% *(3)*.
2. Centrifuge (2500g for 5 min) to pellet the dicyclohexylurea. The faint yellow supernatant contains THEO-O-PNP. Assuming a 60% yield, the THEO-O-PNP concentration is 200 mM.

3.1.6.2. BPE-THEOPHYLLINE (BPE-THEO)

1. Prepare the BPE as in **Subheading 2.2.1.3.** Measure the concentration from the molar absorptivity, 2.41×10^6 AL/mol/cm at 540 nm. Adjust the concentration of BPE to 2.0 mg/mL in PBS, pH 7.4.
2. To 1.0 mL gently stirring BPE in PBS, add 200 μL THEO-O-PNP product in DMF, diluted to achieve >4000-fold molar excess.
3. Protect from light and allow to react overnight at 4°C.
4. Dialyze the product, BPE-THEO, three times against PBS (1000-fold excess) at 4°C and determine the BPE concentration using its molar absorptivity.

3.1.7. Antitheophylline Antibody

Select the theophylline antibody with a dissociation rate sufficiently fast for use in reversible immunosensors (*see* **Subheading 3.2.**). Four of the anti–theophylline antibodies we investigated had dissociation rate constants between 1×10^{-3} and 8×10^{-4} s^{-1} *(3)* with equilibration times of about 50 min in solution, or 100 min when in an intact sensor. Prepare the antibody with affinity chromatography as in **Subheading 3.1.2.1.** For biotinylation, carry out the procedure outlined in **Subheading 3.1.2.2.**, changing the molar ratio of biotin-X-NHS to BPE to 20:1.

3.1.8. Final Reagent Assembly

Reagent concentrations for theophylline sensors are similar to those for PHT: approx 20 nM BPE-THEO, 60 nM biotinylated antibody, and 160 nM Texas Red–streptavidin.

3.2. Antibody Selection and Initial Evaluation

3.2.1. Selection of Antibody

Monoclonal antibodies can be obtained commercially or produced in-house; however, it is of fundamental importance that the selected antibody recognize the epitope exposed after attachment of the antigen to BPE. Therefore, screen only antibodies that were raised using the same functional group for attachment to the carrier as was used to attach to BPE.

3.2.2. Measuring the Dissociation Rate Constant

Simulation experiments suggest that a dissociation rate constant on the order of 10^{-3} s^{-1} will be sufficiently fast to produce a sensor with a response time of < 0.5 h.

1. Measure the dissociation rate constant by adding excess analyte in a small volume to the stirred reagents and follow the fluorescence signal over time with either a standard fluorometer or the laser fiber fluorometer. It is important to add the analyte to the reagents in >1000-fold molar excess so that, once dissociated, antibody is very unlikely to rebind to BPE-antigen.
2. For each measurement, calculate the difference in the maximum fluorescence in the presence of excess antigen and the fluorescence ($F_{max} - Ft$) at time, t. Plot the natural logarithm of the fluorescence change, $F_{max} - Ft$, vs time as shown in **Fig. 4** where 20,000-fold molar excess analyte phenytoin was added. A plot of the natural logarithm of the fluorescence change vs time (inset) gives a slope equal to the dissociation rate constant; 4×10^{-3} s^{-1} in this example *(11)*.

3.3. Reagent Optimization

3.3.1. Antibody Labeling

Once the appropriate antibodies are obtained for evaluation, it will be necessary to optimize the labeling conditions for each clone. As is true for measuring the dissociation rate constant, the simplest approach is to use either a standard fluorometer or the laser fiber fluorometer rather than constructing intact sensors. In the latter case measurements are made using bare fibers placed in a stirred solution.

It is important to achieve sufficient biotin labeling; however, even the relatively gentle process of biotinylating antibody with biotin-X-NHS can result in

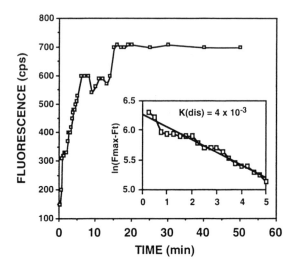

Fig. 4. Measuring the dissociation rate constant of TR-antibody from BPE-PHT. Free phenytoin (400 μ*M* in reaction volume) was added to 20 n*M* BPE-PHT plus 200 n*M* TR-antibody and fluorescence was measured with the fiberoptic fluorometer. The slope of the natural log of the difference between maximum fluorescence and fluorescence at time *t* plotted vs time yields k_{dis} (inset).

denaturation. Therefore, some optimal ratio of biotinylating reagent to antibody exists for each clone, and because there is pH-dependent hydrolysis of NHS, the exact conditions are best determined empirically.

1. For each clone, carry out biotinylation at pH 9.0. Use several widely spaced biotin-X-NHS to antibody ratios. Vary the stock biotin-X-NHS concentration while holding the volume of DMF to antibody solution constant at 5% or less (DMF at greater concentration may denature the antibody).
2. Test the effectiveness of biotinylation by titrating BPE-PHT with antibody in the presence of excess TR-streptavidin using a standard fluorometer or the fiberoptic fluorometer. To 10 n*M* BPE-PHT or other antigen add 0.8–1.0 μ*M* TR-avidin to allow maximal quenching of BPE.
3. Add 10 n*M* aliquots of antibody, allowing at least 15 min for equilibration.
4. Record fluorescence and correct for dilution.
5. Graph the results and interpret as shown in **Fig. 5.** In this example, a 25:1 molar ratio of biotin-X-NHS to antibody was insufficient to accommodate maximum attachment of TR-avidin. A ratio of 50:1 was not significantly different from 75:1. Presumably, under these conditions little additional biotin could be loaded onto the antibody with molar ratios exceeding 50:1. It was found in a separate experiment that using too much biotinylating reagent inhibited the ability of antibody to quench, and a ratio of 50:1 was judged to be optimal.

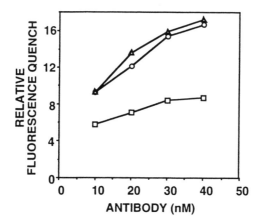

Fig. 5. Optimizing conditions for antibody biotinylation. Purified antibody was split into three fractions that were treated with 25:1 -□-; 50:1 -○-; or 75:1 -△-molar ratios of biotin-X-NHS. BPE-PHT, 10 nM was mixed with 800 nM TR-avidin in PBS, and aliquots of antibody were added in 10-nM increments.

3.3.2. Evaluating the Reagent System

Fluorescence quenching and reversal are both necessary conditions for a working sensor.

1. First ensure that labeled antibody quenches fluorescence from BPE-antigen in the absence of free analyte antigen. This can be done stepwise as in **Subheading 3.3.1.**, or in one step by adding excess antibody (100 nM) and TR-avidin (1 μM) to 10 nM BPE-antigen. Typically, addition of saturating TR-antibody results in about a 30% quench of BPE-antigen fluorescence. Add TR-avidin before antibody to ensure that there is minimal nonspecific quenching. If there is insufficient antigen labeling of BPE, or if the antigen is not available for antibody binding, there will be poor quenching.
2. Optimize the analytical sensitivity by maximizing the transfer of energy between Texas Red and antibody. Generally, quench improves with more efficient labeling of streptavidin with Texas Red (although *see* **Note 3c**). Ensure that TR-avidin exceeds antibody by at least five- to sevenfold (molar) to maximize the transfer of energy when binding occurs. Texas Red is preferred because of its hydrophilicity, but if other labels are tried they should have an absorption spectrum close to the emission spectrum of BPE.
3. Ensure that denaturation of the antibody by the biotinylation process (or by direct labeling) is not causing diminished quenching; *see* **Subheading 3.3.1.** to check antibody labeling efficiency.
4. Add a small volume of concentrated antigen to ensure that the quenched fluorescence can be reversed in the range of analytical interest, otherwise there is no

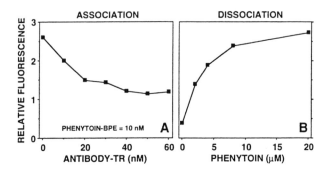

Fig. 6. Demonstration of reagent function with a standard fluorometer. **(A)** The association reaction between TR-antibody and BPE-PHT. Biotinylated antibody was added in sequential aliquots to 10 nM phenytoin-BPE and 1 µM TR-streptavidin in a fluorometer cuvet and allowed to equilibrate. **(B)** In the presence of free PHT, dissociation of TR-antibody from BPE-PHT causes reversal of the quenching. Aliquots of PHT standard in ethanol were added directly to the cuvet.

chance that the reagents will be useful in an intact sensor even if the dissociation rate constant is large. Between 50 and 95% of quenched fluorescence should be reversible with excess analyte *(3,11)*. Ensure that the antigen solvent does not quench.

Figure 6 shows an experiment with a fluorometer demonstrating acceptable quenching and reversal in the presence of analyte. **Figure 6A** shows the association of increasing amounts of TR-antibody with BPE-PHT. Notice that at a TR-antibody to BPE-PHT molar ratio of about 5:1 there is maximal quenching. **Figure 6B** shows the effect of adding increasing amounts of analyte phenytoin to the cuvet, producing a nearly complete reversal of quenching. Under these conditions with a relatively small TR-antibody to BPE-PHT ratio the response is especially steep at low concentrations of analyte. *See* **Subheading 3.5.3.** for notes on adjusting the dynamic range by manipulating the ratio of TR-antibody:BPE-antigen.

3.4. Sensor Construction

Sensors are constructed at the distal end of the fiberoptic using a segment of hollow-fiber dialysis membrane. Refer to **Fig. 7** and **Note 5**.

3.4.1. Fiber Preparation

1. Cleave the optical fiber at 90° with the fiber cleaver.
2. Hold the distal optical fiber in a match flame for a second to burn off the polyimide protective coating. Clean with tissue soaked in acetone.
3. Soak the distal end in fresh Sigmacote™ for 10 min to block free hydroxyl groups. Rinse with DW and air dry.

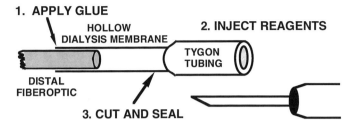

Fig. 7. Assembly of reagents at the distal end of the fiberoptic.

3.4.2. Dialysis Tubing Preparation and Sensor Assembly (*Fig. 7*)

1. Thread 1-cm-long segments of dialysis membrane into half-inch-long segments of Tygon tubing with inner diameter slightly larger than the dialysis tubing and glue in place with rubber cement. Several of these injection guides can be prepared in advance. The Tygon tubing provides a lumen to fill the dialysis tubing with reagents.
2. Load a small volume of reagent into a 1-mL tuberculin syringe with 27-gage needle and purge air.
3. Thread the end of the optical fiber into the dialysis tubing and seal with a drop of cyanoacrylate glue.
4. Immediately insert the syringe into the Tygon tubing and push in a small volume of reagent until a small bleb forms in the glue.
5. Quickly cut the dialysis tubing and seal the end with cyanoacrylate glue.
6. Immediately submerge in gently stirring buffer solution to prevent dehydration.

3.5. Calibration

3.5.1. Standardization

Because sensors measure the concentration of free analyte, it is necessary to use standards that have known concentrations of free analyte. It is also necessary that the standards have an oncotic pressure similar to the test solution. If the test specimen is an aqueous solution, the analyte standards can be prepared in buffer solution. If the test specimen is blood, one alternative is to calibrate with a plasma pool spiked with different concentrations of free analyte. However, because the analyte of interest is likely bound in dynamic equilibrium to albumin or other plasma constituents, this approach requires tedious efforts to measure the concentration of free analyte using a reference method without disturbing the equilibrium. The reader is referred to an example of this approach with phenytoin, which is highly bound to albumin (2).

Whenever possible it will be preferable to prepare primary standards containing known concentrations of analyte in dextran 70 kDa or some other oncotic agent that has been proven not to bind the analyte.

3.5.1.1. Preparation of Primary Standards

1. To test for analyte binding, place the analyte of interest with the oncotic agent in dialysis tubing that will retain the oncotic agent. As a control, put only the analyte in dialysis tubing.
2. Dialyze separately for several hours and assay the dialysate buffer for analyte.
3. If unbound to the oncotic agent, the concentrations of analyte should be the same in both dialysates.

3.5.1.2. Concentration of Standards

1. Select several concentrations of standards (at least five concentrations are recommended) in a range that brackets and exceeds the physiologic (or therapeutic) range. This minimum number is necessary because the dose-response characteristics of a sensor are curvilinear, and because the recommended logit–log transformation makes the lowest and highest points indeterminate. Concentrations of 0, 2, 4, 8, and 20 μM free phenytoin are appropriate because the therapeutic range is 4–8 μM.

3.5.2. Data Reduction

The dose–response behavior follows typical competitive immunoassay characteristics. Several standard data reduction methods can be used for calibration. However, logit–log transformation *(12)* is the suggested data transformation method because it allows signal averaging by linear regression and has the best overall accuracy *(2)*.

1. For each standard, determine $\Delta F/\Delta F_{max}$. Express the difference between the observed fluorescence at equilibrium and the fluorescence at zero concentration (ΔF) as a fraction of the maximum fluorescence increase observed for the highest phenytoin standard concentration (ΔF_{max}). For example, if baseline signal with no analyte was 10,000 counts per second (cps), signal with the most concentrated standard was 13,000 cps, and the signal with a particular standard was 11,200 cps, $\Delta F/\Delta F_{max}$ 1200/3000 = 0.40.
2. Determine logit $\Delta F/\Delta F_{max}$ for each standard. Logit = natural log $[y/(1-y)]$ where y is the fraction. For the above example, logit (0.4) = natural log $0.4/(1-0.4) = -0.405$.
3. Plot log of the measured or nominal free analyte concentration in the standard (x) vs logit $\Delta F/\Delta F_{max}$ (y).
4. Derive the parameters (slope and intercept) for the line using simple linear regression.
5. Determine the $\Delta F/\Delta F_{max}$ for the unknown and use the calibration parameters to estimate the log of the unknown concentration.
6. Obtain the antilog to derive the concentration of the unknown specimen.

3.5.3. Dynamic Range Adjustment

The dynamic range can be adjusted by manipulating the ratio of antibody to BPE-antigen to achieve a slope that is steepest in the range of analytical inter-

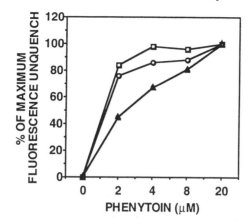

Fig. 8. Effect of antibody to PHT-BPE molar ratio on dynamic range. Analyte phenytoin was added in sequential aliquots to a solution containing 10 nM phenytoin-BPE. Texas Red-labeled antibody was varied to produce molar ratios of 3:1 -□-; 10:1 -○-; and 100:1 -△-. The data from 0–20 μM phenytoin was normalized as $\Delta F/\Delta F_{max}$ (percent).

est as shown in **Fig. 8**. Because the therapeutic range for free phenytoin in blood is 4–8 μM, a 100:1 molar ratio of TR-antibody to phenytoin-BPE was used to affect a dynamic range of 1–20 μM analyte phenytoin response. Increasing the concentration of antibody makes the sensor relatively less sensitive to small concentrations of analyte, consistent with the law of mass action and predictions from simulation experiments *(4)*. Conversely, analytical sensitivity can be increased by decreasing the antibody concentration. The optimum antibody concentration for suitable response over the intended dynamic range may depend on multiple factors and must be determined empirically.

4. Notes

1. Optimizing the optics: The emission monochrometer set at 577 nm can be used to align the optics. Optimal alignment is realized when a constant fluorescence signal from a fluorophore in buffer gives a maximal response, and the background signal from optical reflections is minimized. Fibers with intrinsic fluorescence should not be used; a significant backward reflection can occur at the interface where excitation light is focused onto the fiber, causing a substantial background signal. Background counts should be substantially <1% of the signal from the intact sensor. A reference channel helps to correct for source instability after startup and as a result of fluctuation in ambient temperature.
2. Reagent preparation: PHT-Val can be stored indefinitely if desiccated, but PHT-Val-NHS should be made fresh only after BPE is already prepared for conjugation. Use fresh anhydrous dioxane when preparing PHT-Val-NHS. Water competes in the reaction, producing dicyclohexylurea from DCC. Water is a

likely contaminant if a white precipitate gradually occurs when DCC and NHS are added in dioxane. *See* **Subheading 3.1.1.3.**

3. Labeling efficiency:

    a. The acceptability of BPE-antigen labeling can be checked by titrating a fixed concentration of labeled BPE with increasing amounts of acceptor-labeled antibody (or biotinylated antibody in the presence of excess Texas Red–streptavidin) using a standard fluorometer. Maximal quenching (about 30% after correction for dilution and nonspecific quenching) should occur with an antibody to BPE ratio of about 5 or 6. A significantly lower ratio would suggest that the antigen labeling was inadequate or that there is steric hindrance.

    b. The efficiency of antibody biotinylation can similarly be checked by titrating a fixed amount of BPE-antigen with biotinylated antibody in the presence of excess TR-streptavidin (*see* **Subheadings 3.3.1.** and **3.3.2.**).

    c. Although relatively hydrophillic, Texas Red causes protein aggregation beacuse of fluorophor–fluorophor interactions when too concentrated on protein surfaces. The Texas Red concentration can be estimated directly from its molar absorptivity 8.5×10^4 AL/mol/cm at 596 nm *(9)*. Measuring the streptavidin concentration is more problematic because it cannot be done using A_{280} or with the usual dye binding assays resulting from spectral interference from TR. The streptavidin concentration can be estimated from the nitrogen content as discussed in **Subheading 3.1.3.** It has been suggested that an absorbance ratio of 596 nm/280 nm near 0.8 often gives a useful conjugate *(9)*. If the labeling of streptavidin with Texas Red is sufficient, titration of BPE-antigen in the presence of biotinylated antibody with TR-streptavidin should cause approximately a 30% quench. *See also* **Subheading 3.3.2.**

4. Tendency to precipitation with multiple labeling: If direct antibody labeling is not possible, and an indirect biotin–streptavidin link must be used, pretreatment of TR-streptavidin with a 3:1 molar ratio of free D(+)biotin (Calbiochem, La Jolla, CA) may limit formation of complexes *(11)*. Make a 12-fold concentrated solution of biotin and add 1 part to 3 parts of TR-streptavidin to yield a 3:1 ratio of biotin to streptavidin.

5. Sensor construction: The prepared fiber end should be supported vertically for easiest assembly. Work quickly and submerge the sensor immediately after assembly. If a bubble appears at the interface between the distal fiber surface and the reagents the sensor will not work. Degassing under vacuum or using solutions that have equilibrated at ambient temperature may help prevent the formation of bubbles inside the sensor.

6. When making readings the sensor reagents should be illuminated for only a few seconds to minimize photobleaching. At other times the incident light should be blocked with a shutter.

7. Check the effect of pH, oncotic pressure, ionic strength, and temperature on baseline signal. Fluctuations in pH may cause unpredictable changes in the signal caused perhaps by changes in absorbance of the donor or acceptor, fluorescence quenching, or changes in affinity as a result of altered ionization of the

analyte or antibody-binding site *(2)*. These parameters must be matched between the calibrators and test solution unless they are demonstrated to have no effect. *See* **Subheading 3.1.4.**

Disclaimer

Use of trade names and commercial sources is for identification only and does not imply endorsement by the US Department of Health and Human Services or by the Public Health Service.

References

1. Anderson, F. P. and Miller, W. G. (1988) Fiber optic immunochemical sensor for continuous, reversible measurement of phenytoin. *Clin. Chem.* **34**, 1417–1421.
2. Astles, J. R. and Miller, W. G. (1994) Measurement of free phenytoin in blood with a self-contained fiber-optic immunosensor. *Anal. Chem.* **66**, 1675–1682.
3. Hanbury, C. M., Miller, W. G., and Harris, R. B. (1996) Antibody characteristics for a continuous response fiber optic immunosensor for theophylline. *Biosens. Bioelectron.* **11**, 1129–1138.
4. Miller, W. G. and Anderson, F. P. (1989) Antibody properties for chemically reversible biosensor applications. *Clin. Chim. Acta.* **227**, 135–143.
5. Pecht, I. (1982) Dynamic aspects of antibody function, in *The Antigens* (Sela, M., ed.), Academic, New York, pp. 1–68.
6. Cook, C. E., Kepler, J. A., and Christensen, H. D. (1973) Antiserum to diphenyl-hydantoin: preparation and characterization. *Res. Commun. Chem. Pathol. Pharmacol.* **5**, 767–774.
7. Kronick, M. N. and Grossman, P. D. (1983) Immunoassay techniques with fluorescent phycobiliprotein conjugates. *Clin. Chem.* **29**, 1582–1586.
8. Schultze, H. E. and Heremans J. F., eds. (1966) Survey of plasma proteins, in *Molecular Biology of Human Proteins, vol. 1: Nature and Metabolism of Extracellular Proteins,* Elsevier, Amsterdam, p. 222.
9. Titus, J. A., Haugland, R., Sharrow, S. O., and Segal, D. M. (1982) Texas Red, a hydrophilic, red-emitting fluorophore for use with fluorescein in dual parameter flow microfluorometric and fluorescence microscopic studies. *J. Immunol. Meth.* **50**, 193–204.
10. Haurowitz, F. (1963) Purification, isolation, and determination of proteins, in *The Chemistry and Function of Proteins*, 2nd ed. (Horowitz, F., ed.), Academic, New York, p. 20.
11. Astles, J. R. and Miller, W. G. (1993) Reversible fiber-optic immunosensor measurements. *Sens. Act.* **B11**, 73–78.
12. Thompson, S. G. (1989) Competitive binding assays, in *Clinical Chemistry. Theory, Analysis and Correlation* (Kaplan, L. A. and Pesce, A. J., eds.), Mosby, St. Louis, MO, p. 191–212.

8

Immunobiosensors Based on Grating Couplers

Ursula Bilitewski, Frank Bier, and Albrecht Brandenburg

1. Introduction

Immunosensors based on grating couplers belong to the group of direct optical affinity sensors *(1)*. They allow label-free monitoring not only of immunoaffinity reactions, i.e., of antigen (hapten)–antibody-binding *(2–8)*, but also of receptor–ligand- *(9)*, protein–lipid *(10,11)*, and protein–DNA *(12,13)* interactions, and DNA hybridization *(13)*. In each case one of the binding partners is immobilized on the surface of the optical waveguide and binding of the other partner, present in solution, is monitored. Thus, grating coupler systems allow real-time monitoring of the binding reaction and consequently evaluation of kinetic *(14–16)* and thermodynamic data *(4,12)*. These are common features with other direct affinity sensor systems, such as surface plasmon resonance sensors *(15)*, resonant mirrors *(17)*, and piezoacoustic transducers *(18)*, which all can be summarized as devices for biomolecular interaction analysis.

In this chapter we briefly introduce the optical principle of integrated optical grating couplers and present methods on how to functionalize the grating coupler surface by suitable immobilization to make it a specific sensing device. We concentrate on an example of general interest, the immobilization of avidin, which can be converted for many different applications by simply adding a biotinylated binder. Some hints on how to modify the procedure for more specific proteins will follow and as a method for the determination of low-mol-wt ligands by inhibition or competitive immunosensing, the immobilization of haptens will be presented. Finally, we discuss how to interpret collected data, and what the conditions should be for deriving kinetic data.

From: *Methods in Biotechnology, Vol. 7: Affinity Biosensors: Techniques and Protocols*
Edited by: K. R. Rogers and A. Mulchandani © Humana Press Inc., Totowa, NJ

waveguiding by total   **field distribution**
internal reflection   **of guided modes**

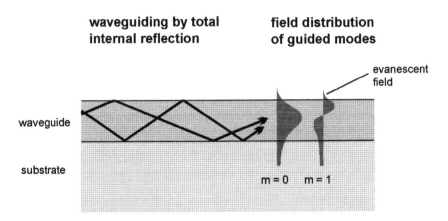

Fig. 1. Light propagation and field distributions in dielectric waveguide.

1.1. Optical Principles of Grating Coupler Systems

Light is guided in a waveguide by multiple reflections at the boundaries (**Fig. 1**). Discrete waveguide modes are formed, which have different field distributions and propagate with specific phase velocities v_p. The velocity of a propagating mode is usually characterized by the effective refractive index n_{eff}, given by:

$$n_{eff} = c/v_p \tag{1}$$

where c denotes the velocity of light in vacuum.

As the electrical field of the waveguide modes penetrates the surrounding media (**Fig. 1**), the propagation of light in the waveguides is influenced by the optical properties not only of the waveguide itself, but also of substrate and cover medium. The part of the field distribution propagating outside the waveguide is called evanescent field because of its strong exponential decay. Changes of the refractive index of the cover medium or of a sensitive film on top of the waveguide cause a change of n_{eff}. The phase velocity is also affected by thickness changes of a sensitive film or by the formation of a layer adherent to the waveguide's surface. The latter influence allows the direct monitoring of biochemical reactions on the surface of a waveguide. One partner of the reaction is immobilized on the waveguide's surface and the binding of the other species is directly observed by detecting the change of the effective refractive index.

Waveguides are optimized to reach a high sensitivity, i.e., a large change of n_{eff} with the adsorption of molecules on the surface *(19)*. For the direct observation of biochemical reactions in aqueous solutions, the best results are achieved with waveguides having a refractive index as high as possible and an optimized film thickness. First experiments have been carried out with SiO_2/TiO_2 wave-

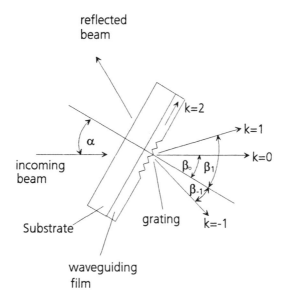

Fig. 2. Light diffraction by a grating coupler.

guides, fabricated by sol-gel technique *(20)*. Besides sensitivity, the stability of the waveguide material, especially the stability of the waveguide's refractive index, is an important aspect. Highly stable waveguides were made of Ta_2O_5 with the ion-plating technique *(21)*. For the case of the sol-gel waveguides, the gratings were formed by embossing. Alternatively, the grating is etched in the substrate's surface, which is structured by means of photolithography.

The change of effective refractive index n_{eff} is determined by the observation of the coupling angle α of a grating coupler *(22)*. The relation between these values is given by the coupling condition:

$$n \sin\alpha = k\lambda_0/\Lambda_0 - n_{eff} \tag{2}$$

where n, k, λ_0, and Λ denote the refractive index in air, the diffraction order, vacuum-wavelength, and grating period, respectively. The geometry of the basic optical configuration is shown in **Fig. 2** with the assumption that the coupling condition is fulfilled for the diffraction order $k = 2$.

Using light sources with small spectral-line widths, such as lasers, the range of angles α at which coupling occurs, is small. Monitoring the coupling angle, therefore, gives rise to an accurate determination of the effective refractive index. Three optical configurations are proposed: the input, output, and reflection grating coupler.

For operation as an input grating coupler *(3,23)*, a parallel light beam is directed on the grating (**Fig. 3**). By inclining the waveguide sample relative to

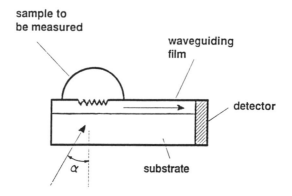

Fig. 3. Input grating coupler.

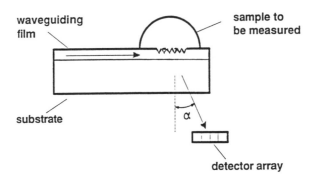

Fig. 4. Output grating coupler.

the direction of the beam, α is changed. If the incoupling condition is fulfilled, a detector positioned at the endface of the waveguide gives a signal. For this mode of operation, the waveguides are easily exchanged.

If the light is launched into the endface of the waveguide, the grating acts as output coupler *(2,5,24)* (**Fig. 4**). In this case, no moving parts are required. The coupling angle is detected, for example, via a position-sensitive detector. On the other hand, this mode of operation demands a critical adjustment at the endface coupling, because the waveguiding films are usually very thin (typically 100–200 nm).

The third way to determine the coupling angle is the reflection grating coupler *(25)*. A range of angles is irradiated onto the grating simultaneously by focusing the incoming laser beam (**Fig. 5**). The reflected intensity distribution is detected. Because of the high efficiency of coupling into guided modes, under the coupling angle a reduced intensity is observed, forming a characteristic minimum in the detected intensity distribution. Adsorptions on the

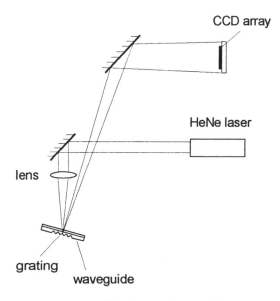

Fig. 5. Reflection grating coupler.

waveguide's surface are monitored by continuously detecting the position of this minimum using a CCD array. An electronic scan of the coupling angle is achieved in combination with a simple way to exchange the waveguides.

The resolution of the system in terms of an effective refractive index is approx $3 \cdot 10^{-6}$ *(25,26)*, which corresponds to a mass coverage of approx 1 ng/cm^2 *(26)*. For analytes with a molecular weight of $10^5 - 1.5 \cdot 10^5$ Daltons, such as hIgG or anti-hIgG, the detection limit is about 1 nM *(23,25)*. However, the sensitivity decreases with the molecular weight of the analyte and molecules with a molecular weight lower than approx 2 kDa at present cannot be directly detected, and a competitive assay format with immobilized haptens has to be chosen. Grating coupler sensors with Ta$_2$O$_5$ waveguides as well as an instrument for monitoring the coupling angle, are provided by Artificial Sensing Instruments (ASI) in Zürich, Switzerland.

1.2. Immobilization of Immunoreagents

A variety of different methods allowing immobilization of proteins or of hapten derivatives taking part in the affinity reaction to be monitored are described in the literature. The final step of the protein immobilization protocol can be performed in the measuring chamber of the grating coupler system, and its efficiency can be monitored on-line, because binding of proteins to the grating coupler during immobilization causes similar binding curves as biochemical affinity reactions. The simplest but often rather effective method for protein immobilization on grating coupler surfaces is physical adsorption. More elaborate and, of course more laborious, are covalent methods often resulting in more stable layers, however, requiring preactivated glass surfaces.

1.2.1. Immobilization Through Adsorption

Physical adsorption of proteins to sensor surfaces is based on van der Waals forces or on ionic interactions. It is generally achieved just by incubation of the sensor with a protein solution *(2,3,7,16)*. However, its effectiveness is strongly dependent on matching properties of the sensor surface and the protein, i.e., their hydrophilicity, hydrophobicity, or surface charge. On the surface of unmodified grating couplers, metal oxide and hydroxide groups are present because of the chemical composition of the waveguide (SiO_2/TiO_2 or Ta_2O_5). These groups may be partly dissociated and charged *(12)* depending on the pH of the protein solution and the effectiveness of protein adsorption is different for different proteins *(2,3,16)*. Thus, a strong adsorption of avidin to completely untreated Ta_2O_5 waveguides was observed whereas adsorption of protein G was only poor *(16)*. Depending on the pI of the protein and the pH of the protein solution, strong binding of the protein as a result of electrostatic binding can be achieved *(16)*. However, the resulting protein layers may be liable to changes of pH and ionic strength of the sample solution and during regeneration cycles. They may be stabilized by crosslinking within the protein layer (**Subheading 3., steps 3–7**).

The waveguide surface properties can be significantly changed by modification with silanizing reagents, introducing amino groups, or with carboxymethyl-dextran layers, introducing carboxylic groups. The result of these modifications of the surface is significantly reduced adsorption of many proteins, and they are the basis for covalent coupling of specific binders.

1.2.2. Immobilization Through Covalent Binding

Covalent binding of proteins to the waveguide surface requires introduction of suitable functional groups, for example amino-, carboxyl-, or epoxy-groups, to which proteins are bound through their amino- or carboxylic groups, either directly or after activation with suitable bifunctional reagents, such as glutaraldehyde *(4)*, bis(sulfosuccinimidyl)suberate (BS^3), or 1-ethyl-3-(3-dimethylaminopropyl)carbodiimide (EDC) (*see* **Subheading 3.4.**). Functional groups may be generated through silanization procedures directly on the waveguide surface (*see* **Subheading 3.3.**) or may be present in a polymer, such as dextran, which is bound on the waveguide *(27)*. Binding to the protein results in Schiff's bases, which can be stabilized by chemical reduction (*see* **Subheading 3.2.**) or in peptide bonds. Binding of hapten derivatives follows a similar protocol as protein immobilization (*see* **Subheading 3.5.**), because those derivatives should be chosen that contain either carboxylic (*see* **Subheading 3.5.**) or amino functionalities *(6)*.

However, there is no general rule for how to achieve an optimum protein layer. Immobilization on a solid phase has an influence on protein structure and thus on its function: Antibodies may preferably bind either at their Fc-part or at the Fab-end near the epitope, which will have great impact on their binding capabilities. Immobilization is therefore crucial for immunosensor performance, and stability and sensitivity depend on it. Also, selectivity of the sensor may be reduced if the surface is not sufficiently covered by specific binding sites, and adsorption of other compounds might interfere with the specific binding.

To avoid time-consuming optimization of immobilization protocols for each ligand, bridges have been introduced, which are binders for other specific receptors. An example of this type is the protein G (or A or a recombinant version A/G), which is a bacterial receptor for the Fc-part of immunoglobulin G (IgG), and facilitates an oriented immobilization of the specific antibodies. Using protein G regeneration of the sensor surface will lead to loss of the specific antibodies, which have to be substituted prior to each measuring cycle.

An even more general approach of immobilizing specific receptors to the surface is the widely used avidin-biotin system. To use this couple, avidin or streptavidin is immobilized on the sensor surface, and a biotinylated receptor, e.g., a biotin-modified antibody, is added. The binding capacity of the antibody should not be influenced by the biotinylation procedure. Usually biotinylation does not lead to an oriented immobilization, since most protocols for biotinylation are not directed to specific sites of the protein. Nevertheless, it has been proven to be an effective method with a broad scope for applications that are not restricted to antibody immobilization. Therefore, in this chapter it was chosen as an example with the methods being transferable to other proteins.

1.3. Data Analysis

In a grating coupler system, binding of high-mol-wt compounds to the sensor surface is monitored in real-time. This feature is independent of the type of grating coupler used (*see* **Subheading 1.**), and in principle the same aspects have to be considered for other direct immunosensing methods. Most of the systems are run as automated flowthrough systems, and in each system a stable baseline is achieved after insertion of the sensor chip into the system and rinsing the chip with a buffer solution. On changing the buffer to the solution to be investigated, binding of a ligand to the sensor surface will take place, dependent on the concentration of ligand. This leads to changes of the refractive index, which consequently changes the coupling angle α (**Eq. 2**), which is recorded as signal. The more ligand binds the more the refractive index changes; thus, a typical binding curve is observed with the binding rate being influenced by the kinetics of the binding reaction. Changing again from sample solution to buffer, ligand that is not tightly bound is removed, and extensive washing should lead finally to a complete removal of ligand. This desorption process can also be monitored through the corresponding changes in refractive index, which finally reaches the original baseline value.

Hence, from the binding curve different parameters can be evaluated. A sufficient incubation time of the protein solution with the sensor leads to an equilibrium between protein binding and protein desorption resulting in a stable sensor signal. The difference from the baseline registered before protein incubation is directly related to the amount of bound protein and can be used for quantitative determination of the analyte. Moreover, for defined affinity reaction partners the binding rate of the protein to the sensor surface is dependent on the protein concentration and thus can also be used for quantitative analysis with reduced analysis times (6). However, binding kinetics are not only influenced by the concentrations of the binding partners but also by the rate constants of the association and the dissociation reactions. Careful data analysis of the time-course of the binding reaction thus allows determination of these kinetic parameters (14) and of derived thermodynamic constants, comparable to other direct sensing methods, such as surface plasmon resonance (SPR). There is still an ongoing discussion about the value of affinity constants determined by these methods, because in contrast to other methods one of the binding partners is immobilized. However, many studies have verified a good correlation of the thermodynamic data, whereas in other reports a significant disagreement was found for special binding pairs.

The immobilization of one partner of the binding reaction has some consequences on the experimental design. There are mass transport effects related to the transport of ligands to the immobilized binder on the sensor surface. Glaser (28) calculated mass transport for planar sensor surfaces; the results were applied to SPR, however, they are applicable in general to any planar (sensing) surface modified with fixed binders. The two central findings are: (1) for concentration measurements a high density of binders on the surface is required, although the analyte (ligand) concentration should be low enough to guarantee mass transport limitation, and (2) for the determination of kinetic rate constants, a low density of the binder at the surface should be used combined with a high ligand (analyte) concentration to reduce mass transport effects.

The specific concentrations to be used in an experiment are dependent on the diffusion coefficient D of the ligand and the flow characteristics of the system given by the flow cell geometry (h = height, b = width, l = length) and the flow rate f. As a limiting case, where k_a' converges to k_a, the product of the apparent association rate constant k_a' and the surface loading G (given in mol/m^2) should exceed the Onsager coefficient L of the flow system:

$$k_a' \, G \gg L \tag{3}$$

$$L \approx (D^2 \, f/h^2 \, b \, l)^{1/3} \tag{4}$$

For a protein with a molecular weight of approx 10^5 Daltons the diffusion coefficient D is in the order of magnitude of $5 \cdot 10^{-11}$ m^2/s, resulting for the ASI-grating coupler flow cell with a flow rate of 50 µL/min in an Onsager coefficient L of 10^6 m/s (k_a' should be given in m^3/mol/s).

2. Materials

2.1. Buffers

1. Potassium phosphate buffer (KPP): 100 mM, KH$_2$PO$_4$/K$_2$HPO$_4$, pH 7.5.
2. Phosphate-buffered saline (PBS): 10 mM NaH$_2$PO$_4$/Na$_2$HPO$_4$, 150 mM NaCl, pH 7.5.
3. Tris: 1 M, Tris(hydroxymethyl)aminomethane-HCl, pH 7.5.
4. HEPES: 10 mM, 4-(2-hydroxyethyl)-piperazine-1-ethansulfonic acid, pH 8.0.
5. 0.1% Tween-20 or Triton X-100 as additives to running buffers.

2.2. Cleaning of the Sensor Surface

1. H$_2$SO$_4$ + 10% H$_2$O$_2$ (Piranha solution).
2. 10 M NaOH.
3. Ethanol.

2.3. Adsorption of Avidin

1. 0.1 mg/mL avidin in PBS, pH 7.5.
2. 1% Glutaraldehyde in 0.1 M KPP, pH 7.5.
3. 1 M Tris, pH 7.5.
4. 10 mg/mL NaBH$_4$ in 0.1 mM NaOH.

2.4. Silanization of Sensor Surface

1. Aminopropyltriethoxysilane (APTS) solution: 10% APTS in H$_2$O with HCl to pH 3.45; optional: 5% APTS in dry acetone.

2.5. Binding of Proteins (Antibodies) Through Covalently Immobilized Avidin

1. 0.1 mg/mL Avidin in PBS.
2. 1-Ethyl-3-(3-dimethylaminopropyl)-carbodiimide (EDC).
3. Biotinylamidohexanoic-acid-*N*-hydroxysulfosuccinimide-ester (Sulfo-NHS-biotin).
4. 4-(2-Hydroxyethyl)-piperazine-1-ethansulfonic acid (HEPES) buffer, pH 8.0.
5. Protein solution in HEPES buffer, 1 mg/mL.

2.6. Covalent Immobilization of Haptens or Hapten Derivatives

1. Carboxylated hapten derivative, e.g., dichlorophenoxyacetic acid (2,4-D).
2. N, N'-Dicyclohexylcarbodiimide (DCC).

3. *N*-Hydroxysuccinimde (NHS).
4. Dry dimethylformamide (DMF).

3. Methods
3.1. Preparation of the Sensor Chip

1. Clean with ethanol and distilled H_2O several times.
2. Rinse in piranha solution for at least 5 min.
3. Wash thoroughly in distilled H_2O (minimum 2 min).
4. Dip in 10 *M* NaOH for 30 s (*see* **Note 1**).
5. Transfer to distilled H_2O, avoid air contact, and work on the chip immediately (i.e., either **Subheadings 3.2.** or **3.3.**) (*see* **Note 1**).

3.2. Adsorption of Protein (e.g., Avidin)

1. Transfer the chip to PBS.
2. Add protein solution. The following steps may be useful, if multiple regeneration of the surface is desired:
3. Add glutaraldehyde solution (1 h at room temperature).
4. Saturate with 1 *M* Tris-HCl, pH 7.5 (1 h at room temperature).
5. Reduce Schiff's bases by $NaBH_4$ (10 mg/mL in 0.1 m*M* NaOH) 2 times 30 min, washing with HEPES, pH 8.0, in between.
6. Rinse and store in HEPES, pH 8.0.

Or covalent binding of proteins:

3.3. Silianization with APTS

1. Incubate the sensor chip in 10% APTS, pH 3.45, for at least 2 h at 80°C.
2. Wash in distilled H_2O.
3. Dry at 120°C for at least 1 h.

or alternative procedure:

4. Incubate in 5% APTS in dry acetone for at least 1 h.
5. Wash in dry acetone.
6. Dry at 80°C for at least 1 h.

3.4. Binding of Proteins (Antibodies) Through Covalently Immobilized Avidin (see Notes 2–4)

1. Activate avidin by EDC in buffer.
   a. Dissolve 100 µg avidin in 100 µL PBS.
   b. Dissolve freshly EDC (0.15 *M*) in 500 µL 10 m*M* NaH_2PO_4, pH 4.0.
   c. Add avidin solution and incubate for 2 h at room temperature.
2. Drop 100 µL of activated avidin on the silanized sensor surface and incubate in a wet chamber at room temperature for at least 12 h.
3. Rinse with PBS.

4. Biotinylation of protein (antibody).
   a. Dissolve 1 mg/mL protein in 10 mM HEPES buffer, pH 8.0.
   b. Add sulfo-NHS-biotin in 100-fold excess.
   c. Incubate for 2 h at room temperature.
   d. Dialyze at 4°C overnight against 0.15 M NaCl.
5. Add biotinylated protein (40 nM) to avidin-coated sensor chip and for 20 min.

3.5. Covalent Attachment of Haptens (see Note 5)

This method applies only for haptens or derivatives with carboxylic groups, e.g., peptides or some pesticides.

1. Dissolve 10 μmol hapten, 50 μmol NHS, and 100 μmol DCC in 200 μL DMF (or scale up).
2. Mix and stir gently for 18 h.
3. Remove precipitated urea by centrifugation (2600g, 5 min).
4. Drop 100 μL of activated hapten (supernatant after centrifugation) on a silanized chip and incubate for at least 2 h in a wet chamber.
5. Remove incubation solution and rinse with ethanol and distilled H$_2$O.

3.6. Measurement (see Notes 6–8)

1. Insert chip with immobilized ligand in the flowthrough chamber of the grating coupler instrument.
2. Rinse the chip with PBS containing a detergent until a stable baseline is obtained; typical flow rate is 50–100 μL/min.
3. Change from rinsing PBS containing a detergent to sample solution and incubate the sample in the flowthrough cell for 5–30 min.
4. Change back to buffer and rinse the chip for approx 5 min (*see* **Note 9**).
5. For regeneration rinse the chip with 10 mM HCl, pH 2.0, for 40 s; flow rate can be higher, for example, 1.8 mL/min.
6. Before the next measurement rinse again with PBS containing a detergent to obtain a new baseline; **steps 2–5** are to be repeated for each measurement.

4. Notes

1. **Steps 4** and **5, Subheading 3.1.** require avoiding air contact. This can be achieved either by working under a nitrogen stream or quick experimentators might be able to transfer from the NaOH solution to water without drying out the surface. In this case a big water reservoir should be used.
2. The immobilization protocol described for covalent immobilization of avidin can be transferred without any problem to other proteins, e.g., antibodies.
3. If proteins of a low-mol-wt (approx <10 kDa) are to be biotinylated, the biotinylated form may be less soluble in buffer than the unmodified protein and may precipitate. In this case the excess of sulfo-NHS-biotin has to be reduced.
4. Immobilization of proteins can be followed on-line, if the chip is inserted in the grating coupler system during the incubation with the protein to be immobilized.

Thus, adsorption of avidin or binding of a biotinylated protein to an avidin-covered surface can directly be observed.

5. Sensor chips with immobilized haptens can be dried at room temperature and stored for more than 1 yr without loss of activity.

6. Detergent, such as Tween-20, is added to the rinsing buffer to avoid air bubbles in the flowthrough system, which may stick to the sensor surface.

7. Direct sensing methods (*see also* Chapter 1) are liable to unspecific effects resulting from the binding of proteins to the sensor surface without a biospecific interaction. Thus, the specificity of the signal has to be proven, e.g., by competitive assays.

8. Since signals cannot be predicted from theoretical considerations, the sensor device has to be calibrated before quantitative analysis of unknown samples is possible.

9. Since the refractive index of the solution in contact with the sensor chip influences the signal together with the binding of proteins to the sensor surface, data preferably obtained in the same solutions should be compared, i.e., the baseline obtained in buffer before the incubation of the chip with the sample solution (**Subheading 3.6.**, **step 2**) should be compared with the signals obtained after having switched again to buffer (**Subheading 3.6.**, **step 4**).

10. The analysis time can be shortened by the kinetic approach, i.e., the rate of signal change after addition of the analyte is determined instead of the steady-state signal.

11. If the kinetic data, i.e., the rate of binding, are used for quantitative analysis, care has to be taken about different refractive indices of rinsing buffer and sample solution leading to a rapid change in signal when the sample reaches the chip. The same rapid change with opposite sign is observed when changing from sample back to buffer.

12. Reliable association or dissociation constants of the receptor–ligand pair are only obtained if effects of mass transport on the binding kinetics are excluded, i.e., the ligand is used in excess (*see* **Subheading 1.3.**). This has to be evaluated by careful investigations using different concentration ratios of immobilized protein (receptors) to ligand present in solution.

References

1. Tiefenthaler, K. (1992) Integrated optical couplers as chemical waveguide sensors. *Adv. Biosens.* **2**, 261–289.

2. Lukosz, W., Clerc, D., Nellen, P M., Stamm, C., and Weiss, P. (1991) Output grating couplers on planar optical waveguides as direct immunosensors. *Biosens. Bioelectron.* **6**, 227–232.

3. Nellen, P. M. and Lukosz, W. (1991) Model experiments with integrated optical input grating couplers as direct immunosensors. *Biosens. Bioelectron.* **6**, 517–525.

4. Polzius, R., Bier, F. F., Bilitewski, U., Jäger, V., and Schmid, R. D. (1993) On-line monitoring of monoclonal antibodies in animal cell culture using a grating coupler. *Biotechnol. Bioeng.* **42**, 1287–1292.

5. Clerc, D. and Lukosz, W. (1994) Integrated optical output grating coupler as biochemical sensor. *Sensors Actuators* **B18–19**, 581–586.

6. Bier, F. F. and Schmid, R. D. (1994) Real time analysis of competitive binding using grating coupler immunosensors for pesticide detection. *Biosens. Bioelectron.* **9**, 125–130.
7. Bernard, A. and Bosshard, H. R. (1995) Real-time monitoring of antigen-antibody recognition on a metal oxide surface by an optical grating coupler sensor. *Eur. J. Biochem.* **230**, 416–423.
8. Gao, H., Sänger, M., Luginbühl, R., and Sigrist, H. (1995) Immunosensing with photo-immobilized immunoreagents on planar optical wave guides. *Biosens. Bioelectron.* **10**, 317–328.
9. Jockers, R., Bier, F. F., and Schmid, R. D. (1993) Specific binding of photosynthetic reaction centres to herbicide-modified grating couplers. *Anal. Chim. Acta* **280**, 53–59.
10. Ramsden, J. J., Bachmanova, G. I., and Archakov, A. I. (1996) Immobilization of proteins to lipid bilayers. *Biosens. Bioelectron.* **11**, 523–528.
11. Heyse, S., Vogel, H., Sänger M., and Sigrist, H. (1995) Covalent attachment of functionalized lipid bilayers to planar waveguides for measuring protein binding to biomimetic membranes. *Prot. Sci.* **4**, 2532–2544.
12. Ramsden, J. J. and Dreier, J. (1996) Kinetics of the interaction between DNA and the type IC restriction enzyme *Eco* R124II. *Biochemistry* **35**, 3746–3753.
13. Bier, F. F. and Scheller, F. W. (1996) Label-free observation of DNA-hybridisation and endonuclease activity on a waveguide surface using a grating coupler. *Biosens. Bioelectron.* **11**, 669–674.
14. Ramsden, J. J. (1993) Experimental methods for investigating protein adsorption kinetics at surfaces, *Q. Rev. Biophys.* **27**, 41–105.
15. Lukosz, W. (1991) Principles and sensitivities of integrated optical and surface plasmon sensors for direct affinity sensing and immunosensing. *Biosens. Bioelectron.* **6**, 215–225.
16. Polzius, R., Schneider, T., Bier, F. F., Bilitewski, U., and Koschinski, W. (1996) Optimization of biosensing using grating couplers: immobilization on tantalum oxide waveguides. *Biosens. Bioelectron.* **11**, 503–514.
17. Cush, R., Cronin, J. M., Stewart, J., Maule, C. H., Molloy, J., and Goddard, N. J. (1993) The resonant mirror: a novel optical biosensor for direct sensing of biomolecular interactions part I: principles of operation and associated instrumentation. *Biosens. Bioelectron.* **8**, 347–353.
18. Ngeh-Ngwainbi, J., Suleiman, A., and Guilbault, G. G. (1990) Piezoelectric crystal biosensors. *Biosens. Bioelectron.* **5**, 13–26.
19. Tiefenthaler, K. and Lukosz, W. (1989) Sensitivity of grating couplers as integrated-optical chemical sensors, *J. Opt. Soc. Am.* **B6**, 209–220.
20. Tiefenthaler, K., Briguet, V., Buser, E., Horisberger, M., and Lukosz, W. (1983) Preparation of planar SiO_2-TiO_2 and $LiNbO_3$ waveguides with dip coating method and an embossing technique for fabrication of grating couplers and channel waveguides. *SPIE* **401**, 165–173.
21. Kunz, R. E., Du, C. L., Edlinger, J., Pulker, H. K., and Seifert, M. (1991) Integrated optical sensors based on reactive low-voltage ion-plated films. *Sensors Actuators* **A 25–27**, 155–159.

22. Lukosz, W. and Tiefenthaler, K. (1983) Directional switching in planar wave-guides effected by adsorption-desorption processes. 2nd ECIO, Florence, IEE Conf. Pub. No. 227, London, 152.

23. Nellen, P. M., Tiefenthaler, K., and Lukosz, W. (1990) Integrated optical input grating couplers as chemo- and immunosensors. *Sensors Actuators* **B1,** 592–596.

24. Lukosz, W., Nellen, P., Stamm, C., and Weiss, P. (1990) Output grating couplers on planar waveguides as integrated optical chemical sensors. *Sensors Actuators* **B1,** 585–588.

25. Brandenburg, A., Polzius, R., Bier, F., Bilitewski, U., and Wagner, E. (1996) Direct observation of affinity reactions by reflected mode operation of integrated optical grating coupler. *Sensors Actuators* **B30,** 55–59.

26. Lukosz, W. and Tiefenthaler, K. (1988) Sensitivity of integrated optical grating and prism couplers as (bio)chemical sensors. *Sensors Actuators* **15,** 273–284.

27. Löfas, S. and Johnsson, B. (1990) A novel hydrogel matrix on gold surfaces in surface plasmon resonance sensors for fast and efficient covalent immobilization of ligands. *J. Chem. Soc. Chem. Commun.* **21,** 1526–1528.

28. Glaser, R. W. (1993) Antigen-antibody binding and mass transport by convection and diffusion to a surface: a two-dimensional computer model of binding and dissociation kinetics. *Anal. Biochem.* **213,** 152–161.

9

Receptor Biosensors Based on Optical Detection

Kim R. Rogers and Mohyee E. Eldefrawi

1. Introduction

Neurotransmitter and hormone receptors serve as biosensors for specific chemical signals ranging from low-mol-wt compounds to complex polypeptides. On binding of the target transmitter or hormone, signal amplification and transduction in biologic systems occurs via a variety of mechanisms, ranging from depolarization of neural membrane, G protein-linked synthesis of second messengers, to activation or inhibition of expression of target genes. The combination of these sensitive and specific sensing receptor proteins with electrochemical, optical, and acoustic technologies to form analytical devices is an attractive concept. These receptor-based biosensors could potentially find applications in the medical, diagnostics, food, military, and environmental areas.

Recent reports for receptor-based biosensors have included the use of acetylcholine receptor *(1)*, interleukin-6 receptor *(2)*, major histocompatibility complex-related receptor *(3)*, and amino acid-sensitive receptor-containing crab antennules *(4)*. The majority of these reports, however, have involved the use of the nicotinic acetylcholine receptor (nAChR) from the electric organs of fish *(1,5–9)*. This is most likely because of the large body of information available concerning the elucidation of this receptor's molecular properties and well-known pharmacology, as well as the fact that the acetylcholine receptor can be relatively easily purified in milligram quantities from the electric organ of the electric ray *Torpedo* sp. *(10)*.

The nAChR can be identified and characterized in subcellular preparations by a number of radioactive ligand binding and functional assays. An assay typically used to characterize the skeletal muscle-type nAChR uses radiolabeled α-bungarotoxin (α-BGT), a small polypeptide neurotoxin (isolated

From: *Methods in Biotechnology, Vol. 7: Affinity Biosensors: Techniques and Protocols*
Edited by: K. R. Rogers and A. Mulchandani © Humana Press Inc., Totowa, NJ

from the banded krait, *Bungarus multicictus* venom) that binds with high affinity to the acetylcholine binding site *(11)*.

The basis for the herein described fiberoptic biosensor assay uses fluorescein-labeled α-BGT as a tracer in a competitive assay for various ligands of the receptor. Detection of the binding of this fluorescent probe to receptors that have been immobilized to a quartz fiber is accomplished via a technique using total internal reflectance fluorescence (TIRF). This method allows the sensitive, instantaneous, and continuous detection of binding of ligand tracer to the immobilized receptor.

Signal transduction methods for reported receptor-based biosensors vary considerably, including electrochemical measurements of potential *(4)* and capacitance *(7)*, surface plasmon resonance *(2)*, and optical measurements involving TIRF *(1,8,9)*. Although these transduction methods and receptor assay formats have, to a certain extent, been tailored to specific receptor systems, versatility of many of these transducer systems will allow for the detection of a variety of biologic receptor proteins. For example, the herein described system should be applicable to any receptor that can be immobilized to a quartz surface and be probed using the binding of a fluorescein-labeled ligand.

2. Materials

2.1. Purification of Nicotinic Acetylcholine Receptor (nAChR)

2.1.1. Preparation of Naja α-Neurotoxin Affinity Gel (12)

1. Buffer A: 1 L 2 M Na_2CO_3, pH 11.9.
2. Buffer B: 3 L 0.2 M $NaHCO_3$, pH 9.4.
3. 50 g CNBr.
4. 400 g Sepharose 4B (dry wt) (Sigma, St. Louis, MO).
5. 100 mg *Naja*-α-neurotoxin (Sigma).
6. 100 g Glycine.
7. 0.02% Sodium azide solution (w/v).
8. Acetonitrile.
9. Glassware, beaker (4 L), graduate cylinders, and conical flasks (250 mL capacity).

2.1.2. Extraction and Isolation of nAChR (10)

1. Buffer C: 5 mM Tris-HCl, 0.154 M NaCl, 1 mM ethylenediaminetetra-acetic acid (EDTA), 0.1 mM phenylmethylsulfonyl fluoride (PMSF), pH 7.4.
2. Buffer C + 1% Triton X-100, 100 mL. Add Triton 1% (v/v) slowly to avoid foaming.
3. Buffer C + 10% Triton X-100, 200 mL. Add Triton 10% (v/v) slowly to avoid foaming.

4. 1 M NaCl, 1% Triton X-100, 500 mL. Add Triton 1% (v/v) slowly to avoid foaming.
5. Sharp knife; cutting board.
6. Cheesecloth (10 × 10 in.); rubber band.
7. Four 250-mL centrifuge bottles.
8. Twelve 25-mL capped ultracentrifuge tubes.
9. Waring blender.
10. Scintered glass funnel.
11. Whatman No. 1 filter paper, 6-in. diameter.
12. *Torpedo nobiliana* electric organ (Biofish Associates, Georgetown, MA).
13. Carbamylcholine.
14. Dialysis tubing, 10,000 mol-wt cutoff, 1-in. diameter.
15. 5 mM Tris-HCl, pH 7.2, 20 L.
16. Variable-speed shaker table.
17. Glassware; 100- and 500-mL beakers, 1-L sidearm suction flask, 250-mL conical flasks.

2.1.3. Confirmation of Activity by $^{125}I\alpha$-Bungarotoxin Binding (11)

1. Carboxymethyl cellulose (Whatman CM-52, microgranular, preswollen).
2. Buffer D: 1.0 mM Na$_2$HPO$_4$, 0.01% (v/v) Triton X-100, 0.03% (w/v), sodium azide, pH 7.2.
3. Pasteur pipets, 9-in.
4. Glass wool.
5. ^{125}I-labeled-α-BGT (DuPont NEN, Boston, MA).
6. α-BGT (Sigma).

2.2. Preparation of Fluorescein-Labeled Bungarotoxin (FITC-BGT) (7)

1. 2 mg α-BGT.
2. 1 mg fluorescein isothiocyanate (FITC) on celite (Molecular Probes, Eugene, OR).
3. 50 mM Sodium bicarbonate, pH 9.5.
4. Microcentrifuge.
5. Sephadex G-25 (Sigma).
6. 5 mM Ammonium acetate, pH 5.8.
7. Lyophilizer.
8. Carboxymethyl cellulose CM-52.
9. 0.5 M Ammonium acetate.

2.3. Noncovalent Immobilization of nAChR to Quartz Fibers (7)

1. Phosphate buffer, 10 mM, NaH$_2$PO$_4$, pH 4.0.
2. Purified nAChR (from *Torpedo, see* **Subheading 2.1.1.**).
3. Methanol.
4. Quartz fibers, 1-mm diameter × 60-mm length (Wale Apparatus Co., Hellertown, PA).

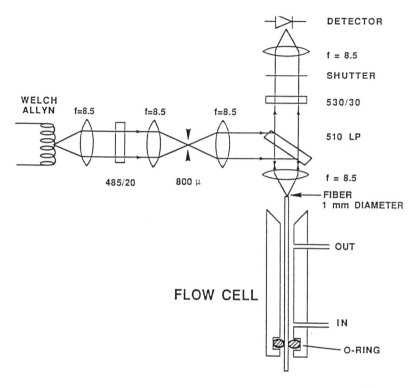

Fig. 1. Schematic presentation of the optical system used to measure fluorescence (with permission from **ref. *1***).

2.4. Biosensor Evaluation (Instrumentation) (7)

1. Evanescent fiberoptic fluorometer equipped with excitation and emission filters at 485/20 nm and 530/30 nm, respectively (ORD Inc, North Salem, NH). Schematics of the optic system are shown in **Fig. 1**.
2. Strip-chart recorder.
3. Peristaltic pump.

3. Methods
3.1. Purification of nAChR
3.1.1. Preparation of Naja α-Neurotoxin Affinity Gel (10)

1. Prepare buffers A and B; keep buffer A at room temperature or it will crystallize; keep buffer B at 4°C.
2. Dissolve 50 g CNBr in 25 mL acetonitrile; mix vigorously (30–60 min).
3. Wash 400 g Sepharose 4B with 100 mL H_2O.
4. Slowly add 800 mL buffer A.

5. Add CNBr solution slowly to the gel slurry, stir vigorously for 2 min at room temperature, then lower temperature to 20°C with ice bath (the CNBr activation is exothermic).
6. Filter gel through scintered glass funnel and wash with 2 L buffer B (0.2 M, NaHCO$_3$).
7. Transfer dry (activated) gel into 500 mL buffer B containing 160 mg *naja*-α-neurotoxin.
8. Divide into 10 flasks and shake overnight at 4°C.
9. Add glycine to a final concentration of 0.5 M and shake 4–5 h at room temperature.
10. Filter gel and wash with H$_2$O.
11. Add 1 L sodium azide solution (0.02%).
12. Divide into 10 aliquots and store at 4°C.

3.1.2. Extraction and Isolation of nAChR (10)

1. Dice 400 g frozen electric organ tissue into 1/4-in. cubes (dicing tissue is most convenient using a large sharp kitchen knife with the tissue partially frozen).
2. Mix 200 g tissue with 500 mL of buffer C and blend in Waring blender for 3 min (slow at first, then increase speed).
3. Repeat with remaining tissue and combine.
4. Let stand 15 min, then pass homogenate through cheesecloth held over the top of a 1-L beaker with a rubber band. One continuous pour is most convenient. Squeeze the liquid from the coarse material in the cheesecloth into the homogenate. Latex gloves should be used for these processes (*see* **Note 1**).
5. Vacuum filter (Whatman No. 1 filter paper) stirring the mixture over the filter to prevent clogging the filter (save a 1-mL aliquot; *see* **Note 5**).
6. Centrifuge filtrate in 250-mL bottles (1000g, 4°C, 10 min, Beckman ultracentrifuge, type-19 rotor or equivalent).
7. Collect pellets into blender using 90 mL buffer C + 1% Triton and mix for 2 min at an intermediate speed; allow 15 min for foaming to subside.
8. Distribute 10 mL homogenate into ultracentrifuge tubes (25-mL polycarbonate), balance, and centrifuge (35,000g, 4°C, 60 min, Beckman ultracentrifuge, type 50.2T rotor or equivalent).
9. Draw off supernatant with Pasteur pipet (save a 1-mL aliquot; *see* **Note 5**); discard pellets.
10. Wash one portion of the *naja*-α-neurotoxin gel (*see* **Subheading 3.1.1.,** **step 12**) with 500 mL H$_2$O and mix filtered gel with supernatant, then divide into two 125-mL Erlenmeyer flasks and shake at 150 rpm for 2 h.
11. Filter mixture through a scintered glass funnel (save a 1-mL aliquot; *see* **Note 5**) and add 100 mL buffer A + 1% Triton; mix well, divide into two 125-mL Erlenmeyer flasks, and shake at 150 rpm at room temperature for 15 min.
12. Repeat **step 11**.
13. Repeat **step 12** with (1 M NaCl + 0.1% Triton).
14. Repeat **step 13**.
16. Mix filtered gel with 50 mL 1 M carbamylcholine, mix well, and shake at 150 rpm at room temperature for 4 h.

17. Filter the gel using a scintered glass funnel.
18. Dialyze filtrate (10,000 mol-wt cutoff dialysis tubing) against 5 mM Tris, pH 7.2, at 4°C. Use of large buffer excess (4 L) and change every 2 h; this will lower the carbamylcholine concentration to $<10^{-10}$ M.
19. Determine the protein concentration by the method of Lowry et al. *(13)* or an alternate method.

3.1.3. Confirmation of Activity by ^{125}I-α-BGT Assay (11)

1. Carboxymethyl cellulose (CM-52) is mixed with 30-fold excess (v/v) buffer D. After several hours, the buffer is decanted. This is repeated until the pH of the decanted buffer remains constant.
2. A glass-wool plug is loosely placed at the bottom of the 9-in. Pasteur pipets and the gel slurry is transferred into columns and allowed to settle to a bed volume of 1.25 mL. Each column is rinsed with buffer B to assure unrestricted liquid flow.
3. The columns are mounted in a side-arm flask using a rubber stopper with a small test tube placed under the end of the pipet column to capture the filtrate. It may be convenient to use tubes that can be directly transferred into the γ counter.
4. The purified nAChR and ^{125}I-α-BGT are drawn through the column (*see* **Note 2**).
5. Reaction mixtures of 125 μL (containing a constant ^{125}I-α-BGT concentration between 1.2×10^{-7} and 9.8×10^{-8} M, and variable nAChR ranging from 0.02 to 0.14 μg protein) are incubated for 1 h.
6. An aliquot (100 μL) is removed from each reaction mixture and loaded onto a gel column followed by immediate filtration and washing with 0.5 mL buffer D.
7. The filtrates are then counted in a γ counter (*see* **Notes 3** and **4**).

3.2. Preparations of Fluorescein-Labeled BGT (FITC-BGT) (7)

1. Add 2 mg α-BGT and 1 mg FITC on celite to 1 mL bicarbonate buffer (50 mM, pH 9.5) and react for 15 min.
2. Centrifuge the reaction mixture (5000g, 5 min) and decant the supernatant to remove the celite.
3. Load the supernatant onto a Sephadex G-25 column (25 × 1.1 cm) and elute with 5 mM ammonium acetate, pH 5.8.
4. Void fractions are pooled, lyophilized, and resuspended in 50 mM ammonium acetate, pH 5.8.
5. Load the sample onto a CM-52 column (10 × 1.5 cm) and elute with the same buffer.
6. After the first peak of fluorescence eluted from the column, the remaining fluorescent material is eluted from the column with 0.5 M ammonium acetate.
7. Fractions that elute with 0.5 M ammonium acetate are pooled, lyophilized, and dissolved in H_2O.
8. The number of fluorescein molecules bound to each α-BGT molecule may be determined by spectrophotometric analysis.

3.3. Noncovalent Immobilization
of the nAChR to the Quartz Fibers (7)

1. Place the quartz fibers in anhydrous methanol for 15 min.
2. Rinse with H_2O, then incubate 30 min in NaH_2PO_4 (10 mM, pH 4.0) containing 50 μg/mL nAChR. This can be efficiently accomplished in 1-mL syringes, single time use, previously washed with methanol.
3. Exchange solution and syringe with PBS buffer and store at 4°C until use.

3.4. Biosensor Evaluation (1)

The binding of FITC-α-BGT to the immobilized nAChR, as well as the effect of agonists and antagonists on this process, can be effectively measured by the inhibition of the initial rate of binding of tracer to the receptor reported by fluorescence. Because of the high affinity of α-BGT for the nAChR, when FITC-α-BGT is used as the tracer the assay is essentially irreversible. The use of lower affinity ligands (such as *naja*-α-neurotoxin) as tracers, however, allows the tracer to be removed from the immobilized receptor by several minutes of buffer perfusion. Indeed, discussion of the variety of assay formats that are possible is beyond the scope of this chapter.

One of the problems universally faced in a binding assay is nonspecific binding. Biosensors may suffer from the same problem. The fluorescent tracer FITC-α-BGT binds nonspecifically to the quartz fiber. Nevertheless, strategies, such as the addition of BSA (0.1 mg/mL) in the assay buffer and reduction of the amount of FITC-α-BGT to low concentrations (i.e., 5 nM), reduce this nonspecific binding to negligible levels. An example of a typical tracing is shown in **Fig. 2**. The low signal-to-noise can be noted in all cases, and the elimination of nonspecific binding using the previously mentioned strategies is shown in **Fig. 2**.

Figure 3 shows the effect of coperfusion of carbamylcholine, *d*-tubocurarine (*d*-TC), and α-BGT on the initial rate of FITC-α-BGT binding to the nAChR-coated fibers. The initial rates of fluorescence signal change can be graphically determined via the strip-chart recorder tracings or numerically determined if electronic data acquisition is used. These rates are normalized as percent maximum response and plotted versus the log of the ligand concentration. The shape and position of these curves are similar, but not identical, to the radioisotope assays. Competition curves for antagonists of the receptor tend to give similar IC_{50} (i.e., the analyte concentration that yields 50% inhibition in tracer binding) values to those measured using radioisotopic methods, whereas agonist curves appear to be shifted about an order of magnitude to the right, i.e., toward lower affinity (*see* **Fig. 3**).

Certain compounds have been shown to change the conformation of the nAChR, resulting in a preincubation-dependent change in binding affinity.

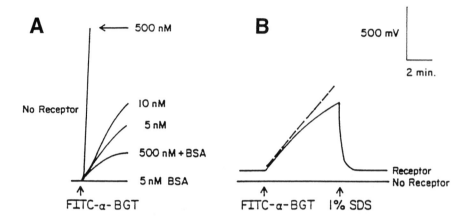

Fig. 2. Binding of fluorescein-labeled α-bungarotoxin (α-BGT) to **(A)** untreated quartz fibers in the presence or absence of bovine serum albinum (BSA) (0.1 mg/mL in phosphate-buffered saline [PBS] buffer), **(B)** untreated or nAChR-coated quartz fibers. Fluorescin isothiocynate-α-BGT (FITC-α-BGT) was introduced at 5 nM in PBS containing BSA (0.1 mg/mL). Dashed line represents graphically determined initial rates. Receptor and FITC-α-BGT were washed from the fiber with 1% sodium dodecyl sulfate (SDS) (with permission from **ref. 1**).

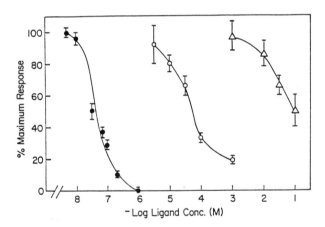

Fig. 3. The effect of various concentrations of *d*-tubocurarine (d-TC) (○), carbamylcholine (△), and α-BGT (●) on binding of 5 nM FITC-α-BGT to nAChR-coated fibers. The various ligands, at the appropriate concentrations, were coperfused with FITC-BGT. There was no pretreatment of the fibers with the ligands. Symbols and bars are means of triplicate measurement ± SEM (with permission from **ref. 1**).

Consequently, one may chose to preincubate the nAChR-coated fibers with the compound of interest or coperfuse the biosensor with the tracer and analyte.

3.5. Discussion

3.5.1. Biosensor Evaluation

The fiberoptic receptor-based biosensor is a generic rapid evanescent wave detector that has been used to measure ligands that bind to receptor-coated optic fibers. In the case of the nAChR, it detects groups of agonists (e.g., acetylcholine, carbamylcholine, nicotine), depolarizing blockers (e.g., decamethonium, succinylcholine), and competitive antagonists (e.g., *d*-TC, pancuronium, *naja*-α-neurotoxin). Consequently, the range of compounds that can be detected depends on the selectivity of the receptor used as the biologic sensing element.

It is important to realize that purification and immobilization of membrane-bound receptors (as in the case of the herein described fiberoptic biosensor) may alter some of the receptor attributes that are observed in their native membrane environment or in solution after purification. It has also been shown, however, that a time-dependent shift in the affinity of nAChR for agonists is still observed when ligand-binding to the free nAChR occurs in solution before the receptor is immobilized, as in the case of an nAChR-based light-addressable potentiometric biosensor *(14)*. Although the fiber optic biosensor can be used as a screening method for various drugs and inhibitors of the nAChR, the caveats related to receptor agonists and different immobilization methods that allow the receptor to undergo free conformational changes must be taken into consideration.

3.5.2. Potential Applications

Biosensor analysis as described above can be applied to any neurotransmitter or hormone receptor that can be isolated in milligram quantities. However, this is not at present a realistic proposition because the nAChR of *Torpedo* is the only receptor that can be harvested in such large quantities. This is because of the presence of this receptor in *Torpedo* electric organs at very high density (0.25 mg receptor protein per gram of tissue). Although affinity chromatography purification protocols have been published for several neurotransmitter receptors, they do not yield sufficient quantities to make biosensor analysis competitive with receptor-ligand binding assays using radioactive ligands.

A new strategy, however, could make biosensor analysis a competitive technology for screening neurotransmitter or hormone receptor ligands. This strategy requires receptor-specific antibodies. Polyclonal and monoclonal antibodies for numerous receptors have been prepared and used as histochemical reagents *(15)*. It would improve the detection capability if antireceptor antibodies were covalently immobilized on the optic fibers using one of several published protocols *(16,17)*. This would result in higher density of antibodies than noncovalent immobilization and the antibodies could serve as anchors for the receptor protein.

Detergent extract of rat brain synaptic membranes can provide a source of soluble receptor proteins. Thus, a 2-h incubation of these antibody-coated fibers in the detergent extract of rat brain synaptic membranes would result in high-affinity binding, so that rinsing the fiber thoroughly in physiologic solution does not dissociate it.

Also, fluorescein-conjugates of receptor antagonists have been synthesized by conjugating any of various fluorescein derivatives (e.g., FITC, fluoresce-inamine, or fluorescein carboxylic acid) to selected receptor antagonists. The chemistry of synthesis of such reagents is not too difficult *(18,19)*. Drugs that bind to a specific receptor displace the fluorescent antagonist of that receptor and reduce fluorescence in a concentration-dependent manner.

4. Notes

1. Phenylmethylsulfonyl fluoride (PMSF) is a potent serine protease inhibitor added to prevent degradation of the nAChR after cell lysis. As a result of the toxicity of buffers and subsequent homogenates containing this compound, appropriate precautions should be taken to prevent ingestion or absorption of these materials.
2. For this assay the CM-52 provides a means of trapping the ^{125}I-α-BGT that is not bound to the nAChR. The efficacy of this column method should be routinely confirmed by measuring ^{125}I-α-BGT that elutes from the gel column in the absence of nAChR. Nonspecific binding of the ^{125}I-α-BGT to the purified nAChR is determined by the addition of 1000 times stoichiometric excess of nonlabeled α-BGT or α-*naja* toxin to the reaction mixture.
3. To circumvent fluctuations in the γ counter operation and counting efficiency, a set of samples of the original ^{125}I-α-BGT stock should be counted with each experiment to determine the specific activity of the toxin.
4. A specific activity of 7 or 8 pmol α-BGT/mg protein is appropriate.
5. It is prudent to save aliquots at various steps in the procedure to assess the increase in specific activity and (if problems in the purification are experienced) to determine at which step in the process they may be occurring.

Notice

The US Environmental Protection Agency (EPA), through its Office of Research and Development (ORD) has, in part, funded the work involved in preparing this chapter. It has been subject to the Agency's peer review and has been approved for publication. Also, the US Army (contracts Nos. DAAM01-940C-0020 and DAA15-89-C-0007) has funded the work involved. The US Government has the right to retain a nonexclusive, royalty-free license in and to any copyright covering this article.

References

1. Rogers, K. R., Valdes, J. J., and Eldefrawi, M. E. (1989) Acetylcholine receptor fiber-optic evanescent fluorosensor. *Anal. Biochem.* **182,** 353–359.

2. Ward, L. D., Howlett, G. J., Hammacher, A., Weinstock, J., Yasukawa, K., Simpson, R. J., and Winzor, D. J. (1995) Use of a biosensor with surface plasmon resonance detection for the determination of binding constants: measurement of interleukin-6 binding to the soluble interleukin-6 receptor. *Biochemistry* **34**, 2901–2907.
3. Raghavan M., Wang, Y. P., and Bjorkman, P. J. (1995) Effects of receptor dimerization on the interaction between the class 1 major histocompatibility complex-related FC receptor and IgG. *Proc. Natl. Acad. Sci. USA* **94**, 11,200–11,204.
4. Buch, R. M. and Rechnitz, G. A. (1989) Intact chemoreceptor-based biosensors: responses and analytical limits. *Biosensors* **4**, 215–230.
5. Eray, M., Dogan, N. S., Reiken, S. R., Sutisna, H., Vanwei, B. J., Koch, A. R., Moffett, D. F., Silber, M., and Davis, W. C. (1995) A highly stable and selective biosensor using modified nicotinic acetylcholine receptor (nAChR). *Biosystems* **35**, 183–188.
6. Nikolelis, D. P., Brennan, J. D., Brown, R. S., McGibbon, G., and Krull, U. J. (1991) Ion permeability through bilayer lipid membranes for biosensor development: control by chemical modification of interfacial regions between phase domains. *Analyst* **116**, 1221–1226.
7. Taylor, R. F., Marenchic, I. G., and Cook, E. J. (1988) An acetylcholine receptor-based biosensor for the detection of cholinergic agents. *Anal. Chim. Acta* **213**, 131–138.
8. Rogers, K. R., Valdes, J. J., and Eldefrawi, M. E. (1991) Effects of receptor concentration, media pH and storage on the nicotinic receptor-transmitted signal in a fiber-optic biosensor. *Biosens. Bioelectron.* **6**, 1–8.
9. Rogers, K. R., Valdes, J. J., Menking, D., Thompson, R., and Eldefrawi M. E. (1991) Pharmacologic specificity of an acetylcholine receptor fiber-optic biosensor. *Biosens. Bioelectron.* **6**, 507–516.
10. Eldefrawi, M. E., and Eldefrawi, A. T. (1973) Purification and molecular properties of the acetylcholine receptor from torpedo electroplax. *Arch. Biochem. Biophys.* **159**, 362–373.
11. Kohanski, R. A., Andrews, J. P., Wins, P., Eldefrawi, M. E., and Hess, G. P. (1977) A simple quantitative assay of 125i-bungarotoxin binding to soluble and membrane-bound acetylcholine receptor protein. *Anal. Biochem.* **80**, 531–539.
12. March, S. C., Parikh, I., and Cuatrecasas, P. (1974) A simplified method for cyanogen bromide activation of agarose for affinity chromatography. *Anal. Biochem.* **60**, 149–152.
13. Lowry, O. H., Rosebrough, N. J., Farr, A. L., and Randall, R. J. (1951) Protein measurement with folin phenol reagent. *J. Biol. Chem.* **193**, 265–275.
14. Rogers, K. R., Fernando, J. C., Thompson, R. J., Valdes, J. J., and Eldefrawi, M. E. (1992) Detection of nicotinic receptor ligands with a light addressable potentiometric sensor. *Anal. Biochem.* **202**, 111–116.
15. Conti-Tronconi, B., Tzartos, S., and Lindstrom, J. (1981) Monoclonal antibodies probes of acetylcholine receptor structure 2. Binding to native receptor. *Biochemistry* **20**, 2181–2191.

16. Bhatia, S. K., Shriver-Lake, L. C., Prior, K. J., Georges, J. H., Calvert, J. M., Bredehorst, R., and Ligler, F. S. (1989) Use of thiol-terminal silanes and heterobifunctional crosslinkers for immobilization of antibodies on silica surfaces. *Anal. Biochem.* **178,** 408–413.

17. Alarie, J. and Sepaniak, M. (1990) Evaluation of antibody immobilization techniques for fiber optic–based fluoroimmunosensing. *Anal. Chim. Acta* **229,** 169–176.

18. Devine, P. J., Anis, N. A., Wright, J., Kim, S., Eldefrawi, A. T., and Eldefrawi, M. E. (1995) A fiber optic cocaine biosensor. *Anal. Biochem.* **227,** 216–224.

19. Colbert, D. L., Gallacher, G., and Mainwaring-Burton, R. W. (1985) Single reagent polarization fluoroimmunoassay for amphetamine in urine. *Clin. Chem.* **31,** 1193–1195.

II

Biosensor-Related Techniques

10

Immunobiosensors Based on Ion-Selective Electrodes

Hanna Radecka and Yoshio Umezawa

1. Introduction

Ion-selective electrode (ISE) methods generally require large volume samples of more than a few milliliters. Therefore, it is highly desirable to devise a simple system for the ISE method to analyze ultratrace amounts of substances in a few microliters of sample solution. This chapter describes a novel and simplified approach for a microliter-ISE system devised by using commercially available conventional ISE without any modification. An essential feature of the present system is the use of a homemade plate-shaped silver/silver halide reference electrode. Containers, such as a beakers, are not needed; instead, a small thin-layer space between a plate reference electrode and the flat bottom of the ISE sensor is conveniently used for holding a few microliters of sample solution. A schematic representation of the principle for the method is shown in **Fig 1**. With this arrangement the necessary sample volume can be reduced to the microliter level (1).

The thin-layer potentiometric system is extremely useful in immunoassay measurements in which the high selectivity without interference from other proteins and use of a sample volume as small as possible are the most important factors. It has been known in biochemistry that the antigen–antibody–complement reaction triggers the formation of "channel-like" holes across the liposome membrane. The combination of this channel-forming phenomenon with ion-selective electrode provides a unique electrochemical immunoassay (2–9).

The principle of the method is as follows. The liposomes are loaded with a concentrated solution of water-soluble membrane-impermeable molecules or ions as a marker. The marker retained within the liposomes will not cause a response in the corresponding ISE. Complement-mediated lysis of liposomes releases the marker ions to a dilute solution, where the relevant ISE can respond sufficiently rapidly

From: *Methods in Biotechnology, Vol. 7: Affinity Biosensors: Techniques and Protocols*
Edited by: K. R. Rogers and A. Mulchandani © Humana Press Inc., Totowa, NJ

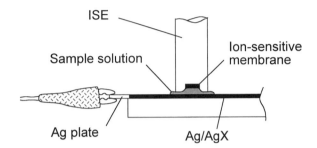

Fig. 1. Schematic diagram of the thin-layer potentiometry.

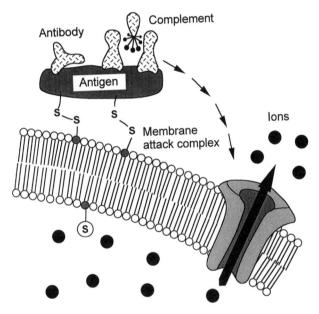

Fig. 2. Schematic diagram of the formation of "channel-like" holes across the liposomes membrane. The antigen–antibody–complement reaction triggers the formation
of channel-like holes that enable entrapped ions to flow through the hole.

under the conditions used. A schematic diagram of the principle of the liposome
immune-lysis process and thin-layer potentiometric assembly is shown in **Figs. 2–4**.

2. Materials

2.1. Fluoride Determination Using
the Plate-Shaped Silver/Silver Chloride Electrode

1. Silver/silver chloride wire.
2. Silver plate (2 × 2 cm, 0.5 mm thick).

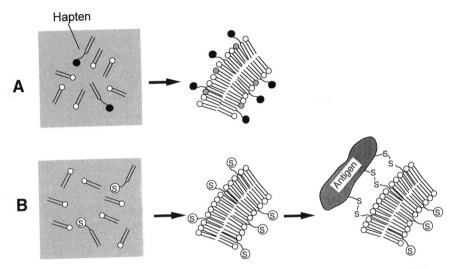

Fig. 3. Immunologic sensitization of the liposome membrane surface by two different methods. (1) Lipid haptens can be liposome constituents by themselves, or (2) protein antigens, which is the general case, are chemically bonded on the membrane surface.

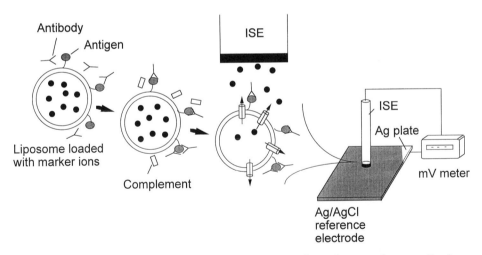

Fig. 4. Release of marker ions through membrane channels on antigen–antibody–complement reaction and after detection of ions in microvolume sample solutions by thin-layer potentiometry.

3. Fluoride standard solution (dissolve an analytical grade NaF in deionized and distilled water).
4. Total ionic strength adjustment buffer: 1 M NaCl, 0.25 M CH$_3$COOH, 0.75 M CH$_3$COONa, 0.001 M sodium citrate.

5. Fluoride ISE.
6. 0.1 M KCl.

2.2. Immunoassay
for an Antihuman IgG Antibody and Human IgG

1. HEPES-buffered salt solution: 8.0 g NaCl, 0.4 g KCl, 0.25 g Na_2HPO_4, 2.38 g 4-(2-hydroxyethyl)-1-piperazineethane sulfonic acid (HEPES); dissolve to 1000 mL of water, pH 7.45, adjusted with 0.1 M NaOH.
2. Potassium fluoride-HEPES-buffered solution: 14.1 g KF · 2 H_2O, 2.38 g HEPES; dissolve to 1000 mL of water, pH 7.45, adjusted with 0.1 M NaOH.
3. Acetate-buffered saline: 8.2 g sodium acetate, 8.5 g NaCl; dissolve to 1000 mL of water, pH 4.50, adjusted with 0.1 M acetic acid.
4. Gelatin veronal-buffered saline (GVB⁻): 10.19 g sodium veronal, 83.0 g NaCl; dissolve to 2000 mL of water, pH 7.40, adjusted with 1 M HCl; for experimental use, dilute the stock solution and supplement with 2% gelatin to final concentration of 0.1% (GVB⁻).
5. Supplement GVB⁻ with 35 mM $MgCl_2$ and 0.15 mM $CaCl_2$ for assay procedures (GVB²⁺).
6. Lipids: cholesterol, dipalmitoylphosphatidylethanolamine (DPPE), dimyristoyl-phosphatidylcholine (DMPC), modified dipalmitoyl-phosphatidylethanolamine (DTP-DPPE).
7. Chloroform.
8. Rabbit antihuman IgG antibody (IgG fraction), purchased from Miles-Yeda (Elkhart, IN).
9. Monoclonal antihuman IgA, IgG, and IgM antibodies, purchased from Diagnostic Technology (Hauppauge, NY).
10. Guinea pig serum (source of complement), stored at –80°C.
11. Specific pathogen-free (SPF) guinea pig complement, purchased from Shizuoka Laboratory Animal Center (Hamamatsu, Japan).
12. N-Hydroxysuccinimidyl 3-(2-pyridylthio)propionate (SPDP) crosslinking reagent.
13. Dithiothreitol (DTT)—reducing reagent.
14. 0.2 M KF.
15. Plate-shaped silver/silver chloride reference electrode (*see* **item 2**, **Subheading 2.1.**).

2.3. Thin-Layer Potentiometric Analysis
of Lipid Antigen–Antibody Reaction
by Tetrapentylammonium (TPA⁺) Ion-Loaded Liposomes

1. Dipalmitoyllecithin, cholesterol, dicetylphosphate, (ε-dinitrophenylamino-caproyl) phosphatidylethanolamine (DNP-cap-PE) as a lipid antigen; molar ratio: 2:1.5:0.2:0.1.
2. Chloroform.
3. Antiserum (anti-DNP).
4. Fresh guinea pig serum (source of complement).
5. 0.15 M tetrapentylammonium chloride (TPA⁺Cl⁻) solution.

6. Modified veronal saline: 3.12 mM barbital, 1.82 mM barbital sodium, 0.15 mM $CaCl_2$, 0.5 mM $MgCl_2$, 0.147 M NaCl; pH is adjusted to 7.4 using HCl.
7. Modified Tris-buffered saline: 17 mM Tris (hydroxylmethyl) amino-methane, 0.15 mM $CaCl_2$, 0.5 mM $MgCl_2$, 0.147 M NaCl; pH is adjusted to 7.4 using HCl.
8. Constituents for TPA$^+$ ISE:
   a. Tetrapentylammonium chloride (TPA$^+$Cl$^-$), 2 mg.
   b. Dioctyl phthalate (DOP),130 mg.
   c. Poly(vinyl chloride), 66 mg.
   d. Tetrahydrofuran (THF), approx 3 mL.
   e. 0.01 M KCl as the internal solution of ISE.
   f. Reference internal electrode (Ag/AgCl wire).

Item 15, Subheading 2.2. is also necessary for this protocol.

2.4. Determination of Anticardiolipin Antibodies in Syphilis Serology

1. Dipalmitoyl phosphatidylcholine (DPPC), cardiolipin (CL), cholesterol (CH), stearylamine (SA), in molar ratio 2:0.02:1.5:0.21.
2. Hapten modified phosphatidylethanolamine (ε-DNP-cap-PE).
3. Chloroform.
4. 0.15 M TPA$^+$Cl$^-$; pH = 7.4.
5. Modified veronal saline (VBS): 3.12 mM barbital, 1.82 mM barbital sodium salt, 0.15 mM $CaCl_2$, 0.5 mM $MgCl_2$, 0.147 M NaCl.
6. Constituents for TPA$^+$Cl$^-$ ISE (*see* **item 8, Subheading 2.3.**).
7. Seropositive and normal human sera (provided by the Japan Red Cross Blood Center, Tokyo).
8. Guinea pig serum (source of complement), stored at –80°C.
9. Fine carbon powder.

Item 15, Subheading 2.2. is also necessary for this protocol.

3. Methods

3.1. Fluoride Determination by Using Plate-Shaped Silver/Silver Chloride Reference Electrode (see Notes 1–3)

1. Anodize a silver plate in 0.1 M KCl at +0.5 V vs saturated calomel electrode (SCE) for approx 5 min.
2. Leave part of the silver plate uncovered by AgCl for electrical contact with a millivolt meter.
3. Place the plate reference electrode in a horizontal position.
4. Drop by a micropipet a few microliters of sample solution onto a plate reference electrode.
5. Lower the ISE body in an upright position onto the plate reference electrode where the sample droplet exists.

6. Close the necessary electrical circuit and measure the potential after a predetermined time to guarantee equilibrium.
7. Pull up the ISE body and rinse both the reference and selective electrodes with deionized and distilled water for subsequent measurements; the sample solution is not stirred during the measurements.

3.2. Immunoassay for an Antihuman IgG and Human IgG Antibody (see Notes 4–9)

1. Preparation of the plate-shaped silver/silver chloride reference electrode (*see* **steps 1** and **2** in **Subheading 3.1.**).
2. Multilamellar liposomes preparation:
   a. Dissolve DMPC, cholesterol, and DTP-DPPE in molar ratio of 1:1:0.06 into $CHCl_3$, using a pear-shaped flask.
   b. Evaporate $CHCl_3$ with rotary evaporator under reduced pressure.
   c. Add 0.2 M KF solution into the dried lipid film.
   d. Incubate at 50°C for 1 min.
   e. Disperse lipid film by vigorous vortexing or sonication (using bath sonicator).
   f. Collect the liposomes by centrifugation at 30,000g for 20 min.
   g. Suspend the pellet liposomes in KF-HEPES buffered solution.
   h. Store the liposomes at 4°C under nitrogen.
3. Preparation of modified human IgG with SPDP and DTT:
   a. Dissolve 5 mg of human IgG in 2 mL of HEPES-buffered salt solution.
   b. Add 0.1 μmol of SPDP under nitrogen gas.
   c. Keep the reaction mixture at room temperature for 30 min.
   d. Charge the mixture on a Sephadex G-25 fine column (10 × 170 mm) pre-equilibrated with saline.
   e. Elute protein-peak fraction (2 mL) with acetate-buffered saline.
   f. Reduce protein with 30 mg of DTT under nitrogen gas.
   g. Incubate 20 min at room temperature.
   h. Apply the mixture to a Sephadex G-25 fine column (15 × 150 mm) pre-equilibrated with HEPES-buffered salt solution.
   i. Store the first main peak fraction (2 mL) at 4°C under nitrogen gas.
4. Preparation of IgG protien-pendant liposomes:
   a. Add 2-mL portion of liposome suspension to the same volume of modified human IgG and react overnight at room temperature with slow shaking.
   b. Remove untrapped marker fluoride ions by repeated centrifugation at 30,000g for 20 min in GVB^{2+}.
   c. Suspend the final pellet of liposomes in 1.2 mL of GVB^{2+} and store at 4°C under nitrogen.
5. Standard assay procedure:
   a. Dilute 25 μL of IgG fraction of rabbit antihuman IgG antibody and 25 μL of complement 50 times with GVB^{2+} and add to 25 μL of liposomes aliquot.
   b. Incubate the mixture for 60 min at 37°C in a moist chamber.
   c. Stop the reaction by cooling at 4°C.

d. Put 25 μL aliquot of the resulting solution on the plate Ag/AgCl reference electrode and lower fluoride ISE onto the sample droplet.

e. Measure the electromotive force (emf.) by a millivolt meter after 1 min when an equilibrium potential is attained.

f. Determine the total releasable F-ions in the liposomes by lysing with 25 μL of a 9% Triton X-100 solution.

g. Estimate the marker release (%) as follows:

$$\text{marker release (\%)} = (\text{experimental release} - \text{blank})/(\text{total release} - \text{blank}) \times 100\% \quad (1)$$

where the blank is obtained by simply replacing the antibody solution with the same volume of GVB^{2+} solution in the above standard assay procedure.

h. Examine the dependence of the extent of the marker ion release on the concentration of antihuman IgG antibody at given amounts of complement and IgG antigen.

6. Assay of human IgG antigen:

a. Incubate a mixture of 25 μL of antihuman IgG antibody (22 μg/mL) and 5 μL of adequately diluted human IgG ($0.8-2.5 \times 10^{-4}$ mg/mL) for 1.5 h at room temperature.

b. Add 25 μL of liposome suspension and 25 μL of 100× diluted complement; incubate for 60 min at 37°C.

c. Measure the marker release (%) degree using plate-shaped Ag/AgCl reference electrode (*see* **steps 5d–g** in **Subheading 3.2.**).

3.3. Thin-Layer Potentiometric Analysis of Lipid Antigen–Antibody Reaction by Tetrapentylammonium (TPA⁺) Ion Loaded Liposomes and TPA⁺ ISE (see Notes 10–15)

1. Preparation of TPA^+ ISE:

a. Dissolve 2 mg of TPA^+ in 130 mg DOP; add approx 3 mL THF.

b. Gradually add 66 mg PVC.

c. Stir the mixture to completely dissolve PVC.

d. Pour out the solution on a Petri dish (3-cm diameter).

e. Cover the dish and allow the THF to slowly evaporate (approx 2 d).

f. Cut a circle of PVC film and fix to a glass tube (5-mm diameter) or to a Denki Kagaku Keiki ISE body (Tokyo, Japan).

g. Use Ag/AgCl wire as the internal reference electrode.

2. Liposomes preparation:

a. Dissolve the mixture of lipids and lipid antigen into $CHCl_3$ in a pear-shaped flask.

b. Evaporate $CHCl_3$ under reduced pressure.

c. Swell the dried lipid films in 0.15 M TPA^+ aqueous solution.

d. Disperse the lipid film by vigorous vortexing or sonication using a bath sonicator.

e. Remove the untrapped marker ions, TPA^+ ions, by centrifugation (15,500g) for 15 min each at 0°C with five changes of the modified veronal saline or modified Tris-buffered saline.

f. Suspend liposomes with 1.2 mL veronal saline or modified Tris-buffered saline and store at 4°C under nitrogen.

3. Typical experimental procedure:

   a. Mix 25 µL (variable) liposome aliquot, 0.5 µL (variable) anti-DNP antise-rum, and 0.5 µL (variable) fresh guinea pig serum; bring this mixture to a total volume of 100 µL with modified Tris-buffered saline.

   b. Incubate for 30 min at room temperature (21 ± 0.5°C).

   c. Drop approx 20–50 µL sample solution onto the plate Ag/AgCl reference electrode (*see* **Subheading 3.1.**).

   d. Measure the potential after 3 min for equilibrium to be attained.

3.4. Determination of Anticardiolipin Antibodies in Syphilis Serology (see Notes 16–20)

1. Preparation of liposomes (*see* **step 2** in **Subheading 3.3.**).

2. Preparation of plate-shaped Ag/AgCl reference electrode (*see* **Subheading 3.1.**).

3. Preparation of TPA$^+$ ISE (*see* **step 1** in **Subheading 3.3.**).

4. Determination of the titer value with a standard semiquantitative method SST (Serodiagnosis of Syphilis Test, Iatron Co., Tokyo, Japan). The SST method is based on careful observation by the naked eye of the extent of precipitates composed of antigen–antibody complexes and fine carbon powder. The reagents used for the SST method are the same lipids as those for the con-stituents of liposomes (*see* **item 1**, **Subheading 2.4.**), except for stearylamine (which is added to increase the trapping efficiency of liposomes for posi-tively charged marker ions). The titer value for the seropositive sera used here is 1.512.

5. The typical procedure:

   a. Add to a 50 µL or other (suitable) liposome aliquot an equal volume of Wassermann-seropositive serum and fresh guinea pig serum.

   b. Incubate at 25°C for 30 min.

   c. Place approx 20–50 µL of mixture on the plate Ag/AgCl reference electrode.

   d. Measure the potential after 3 min for equilibrium to be attained (*see* **Sub-heading 3.1.**).

   e. Heat the liposomes at 56°C for 30 min (to inactivate complement) and mea-sure the emf for inactive complement liposomes.

   f. Express the results as a $\Delta E = E_c - E_c'$; where E_c is the emf value with active complement and E_c' is the emf value with deactivated complement.

6. Minimization of anticomplement reaction induced by unidentified constituents in human sera:

   a. Prepare the haptenic antigen-sensitized liposomes using the lipids (*see* **item 1**, **Subheading 2.4.**); replace cardiolipin with DNP-cap-PE.

   b. Heat the antigen-sensitized liposomes with anitserum at 37°C for 30 min.

   c. Wash adequately with VBS by centrifugation (30,000*g*) with two changes of VBS.

   d. Examine the dependence of the extent of the marker-ion release on the con-centration of complement at given amount of antiserum.

   e. Measure the potential according to **step 5a–f** in **Subheading 3.4.**

4. Notes

4.1. Fluoride Determination by Plate-Shaped Silver/Silver Chloride Reference Electrode

1. It is important to note that although the Ag/AgCl electrode is by itself a chloride ion-selective electrode, unwanted shifts of the reference potential caused by possible variation of chloride ion activity from one sample to another can be eliminated by the use of saline buffers. It should also be pointed out that the influence of serum proteins on the potential change of the Ag/AgCl reference electrode is negligible, even though the latter is in direct contact with the sample solution. Because of these advantages, the present approach has been conveniently used for a potentiometric immunosensor combined with liposome immunochemistry.

2. For common ISEs used, a polyethylene layer surrounds the solid or liquid membrane and protrudes (about 0.2 mm) from the electrode surface. This type of ISE design prevents unwanted direct electrical short circuit of the sensitive membrane with the plate reference electrode.

3. The necessary solution volume is essentially dependent on the structure of the bottom of ISE; it could be as low as 3 μL. When the sample volume was decreased to 1 μL, the corresponding concentration observed was slightly higher than the actual one. This is probably because the effect of sample evaporation is no longer negligible.

4.2. Immunoassay for an Antihuman IgG and Human IgG Antibody

4. The estimation of the amount of pendant IgG was carried out by measuring proteins in the supernatant before and after the coupling reaction. For the determination, the commercially available protein assay kit can be used. Coomassie Brilliant blue G-250 is applied as a staining dye.

5. The reaction of SPDP with molecules containing primary amino groups results in the introduction of dithiopyridyl groups through the formation of amide bonds and release of N-hydroxysuccinimide. Protein-bound dithiopyridine (DTP) can be readily activated as the thiol derivative by reduction with dithiothreitol (DTT) in mild conditions (acidic pH). These thiolated proteins can react with other DTP-substituted molecules to produce covalently coupled products. For coupling of proteins to liposomes, SPDP reacts with DPPE in organic solvent to form the stable derivative DTP-DPPE. This compound can be mixed with other lipids and form liposomes after sonication. Human IgG modified with SPDP is used for coupling to the liposomes. In the coupling reaction with crosslinking reagent (SPDP) and reducing agent (DTT), normal S–S bonds in protein molecules are not reduced. Therefore, the affinity of modified human IgG against antibody is not changed.

6. The marker-ion release increases with increasing the complement concentration. However, too high a complement concentration seems to cause nonspecific changes in the permeation rate of marker ions through the liposome membrane. Thus, the 100× dilution of complement is employed as a complement source.

7. At a given amount of complement and IgG antigen, marker release starts to occur when antibody concentration becomes $>2 \times 10^{-3}$ mg/mL and gradually levels off at about 30% marker release above 2×10^{-1} mg/mL of antibody; thus, one can determine the antihuman IgG antibody level of 2×10^{-3} through 2×10^{-1} mg/mL.

8. Human IgG (as protein antigen) pendant on the liposomes is virtually specific to antihuman IgG antibody, when the concentration of antihuman IgA and IgM antibodies are both at 10^2 dilution or less, so that monoclonal antihuman IgA and IgM antibodies do not interfere with this determination of antihuman IgG antibody in this concentration range.

9. Inhibition of the immune reaction also occurs when a free antigen that has crossreactivity with the corresponding antibody coexists in solution. In the present protocol, the minimum amount of antihuman IgG antibody that is necessary for the maximum channel formation of liposomes is first reacted with a known amount of free human IgG in aqueous solution. The degree of inhibition of antibody activity caused by this reaction is then measured by adding the liposome aliquot. The antibody activity that remained after reaction with the free human IgG is measured by the degree of F^- release from added liposomes in the presence of complement. The degree of inhibition thus measured is dependent on free (solution) antigen concentration and this fact can be used for the assay for human IgG antigen.

4.3. Thin-Layer Potentiometric Analysis of Lipid Antigen-Antibody Reaction by Tetrapentylammonium (TPA⁺) Ion Loaded Liposomes and TPA⁺ ISE

10. Longer storage (approx >12 h) of the liposome aliquot leads to precipitation of liposomes and seems to cause scattered data for unknown reason. Therefore, the liposome aliquot is prepared fresh before each experiment.

11. TPA$^+$ is chosen as a marker ion in order to minimize the background leakage from the liposomes.

12. TPA$^+$ ISE exhibits a Nernstian response to TPA$^+$ from 0.1 to 5×10^{-6} M. The selectivity coefficient against Na$^+$, K$^+$, and Li$^+$ are 5.0×10^{-6}, 6.2×10^{-6}, and 6.0×10^{-7}, respectively. Therefore, electrode interference from Na$^+$ and K$^+$ ions in serum is negligible.

13. Sample volume can be decreased to 8 μL.

14. In the presence of an optimal amount of guinea pig serum as a complement source, immune lysis of liposomes sensitized with DNP-cap-PE is induced by serum containing specific antibodies. The observed potential change on changing the concentration of antiserum is caused only by the corresponding immune reaction rather than the change of ionic strength, pH, and the concentration of serum proteins that are nonspecific to relevant immune reactions. Immune lysis of liposomes sensitized with DNP-cap-PE is also dependent on the amount of complement. Thus, one can determine either the anti-DNP antibody or complement levels by the present method.

15. When the complement is deactivated by heating at 56°C for 30 min, the immune lysis reaction does not occur even with increasing complement levels, and therefore no potential change is observed. This supports the well-established principle that immune lysis of liposomes is caused by the characteristic biochemical action of complement.

4.4. The Determination of Anticardiolipin Antibodies in Syphilis Serology

16. Liposomes prepared from the mixture of the pure lipids, cholesterol, lecithin, and cardiolipin are able to bind antibodies against the spirochete causing the disease (*Treponema pallidum*). In the presence of complement, an immunologic lysis of the membrane takes place (**Figs. 2–4**).

17. All human sera used are treated at 56°C for 30 min before testing in order to eliminate the complement activity from human serum itself. The complement used is from guinea pig serum, which is stored at –80°C. The titer value for this is 267 CH50/mL by a simplified Mayer method.

18. It is known that the phase-transition temperature for liposomes differs according to the kinds of phospholipids used. Thus, it seems important to choose an appropriate temperature for incubation. We selected an incubation temperature of 25°C for the immune lysis reaction as a tradeoff between the greater possibility of background release of marker ions at higher temperatures and the optimum temperature of around 37°C for the immune reaction.

19. The crucial point for obtaining maximum potential change is minimization of the anticomplement reaction induced by unidentified constituents in human sera. To do this, the haptenic antigen-sensitized liposomes are first complexed with corresponding antibody (Wassermann antibody) and any unwanted components in the sera are separated from the system.

20. The results are expressed as a difference between the emf reading when the active complement was used and the emf when the inactive complement (heat-treated at 56°C for 30 min) was used but under otherwise identical conditions. The advantage of using ΔE is that the background correction is accurate because the total protein concentration is the same for the sample (with active complement) and the blank solution (with deactivated complement), even if the level of complement has to be changed.

References

1. Chiba, K., Tsunoda, K., Umezawa, Y., Haraguchi, H., Fujiwara, S., and Fuwa, K. (1980) Plate-shaped silver/silver halide determination of fluoride ion in microliter solution with fluoride ion selective electrode. *Anal. Chem.* **52,** 596–598.
2. Abe, H., Kataoka, M., Yasuda, T., and Umezawa, Y. (1986) Immunoassay using ion selective electrode and protein pendant liposomes. *Anal. Sci.* **2,** 523–527.
3. Shiba, K., Umezawa, Y., Watanabe, T., Ogawa, S., and Fujiwara, S. (1980) Thin-layer potentiometric analysis of lipid antigen-antibody reaction by tetrapentyl-ammonium (TPA$^+$) ion loaded liposomes and TPA$^+$ ion selective electrode. *Anal. Chem.* **52,** 1610–1613.

4. Umezawa, Y., Sofue, S., and Takamoto, Y. (1984) Thin-layer ion selective electrode detection of anticardiolipin antibodies in syphilis serology. *Talanta* **31,** 375–378.
5. Umezawa, Y., Kataoka, M., Sugawara, M., Abe, H., Kojima, M., Takinami, M., Sazawa, H., and Yasuda, Y. (1987) Immunosensor systems using liposomes and planar lipid bilayer membranes for ion-channel model sensors (Schmid, R. D., ed.), Biosensors International Workshop 1987, GBF Monograph, vol. 10, pp. 139–144.
6. Umezawa, Y. (1983) Ion-selective immunoelectrode, in *Proceedings of the International Meeting on Chemical Sensors* (Seiyama, T., Fueki, K., Shiokawa, J., and Suzuki, S., eds.) Elsevier, Fukuoka, Japan, pp. 705–710.
7. Shiba, K., Watanabe, T., Umezawa, Y., Fujiwara, S., and Momoi, H. (1980) Liposome immunoelectrode. *Chem. Lett.* 155–158.
8. Umezawa, Y. and Sugawara, M. (1988) Ion sensors for microsampling, in *Chemical Sensor Technology*, vol. 1 (Seiyama, T., ed.), Kodansha Ltd., Tokyo, Japan, pp. 141–152.
9. Umezawa, Y. and Fujiwara, S. (1980) Thin-layer potentiometry. *Nippon Kagaku Kaishi* 1437–1441.

11

Biosensors Based
on DNA Intercalation Using Light Polarization

John J. Horvath

1. Introduction

The intercalation of polyaromatic compounds by DNA can serve as a basis for a simple and sensitive method for detection and quantification of carcinogens. The experimental technique is based on monitoring the decrease of polarization, caused by the displacement of an intercalated fluorescent dye molecule by the analyte molecule (carcinogen). The magnitude of the polarization decrease is proportional to the concentration of the analyte.

Intercalation is a reversible insertion of a guest species into a lamellar host structure. Study of the reactions between guest molecules and the host molecule (double-stranded DNA) has been ongoing since 1947, when Michaelis (1) observed and correlated dramatic changes in the visible absorption spectra of basic dyes when binding to DNA. Quantitative binding studies have been made by using equilibrium dialysis (2,3), thermodynamic models, such as Scatchard plots (4), viscosity (5), NMR (6), and fluorescence spectroscopy (7–10).

The intercalative interactions of dyes with DNA have been intensively studied and characterized by using many different methods (11–18). In addition to dyes, other compounds, such as aminoquinolines (19), fused aromatics, such as diamino-phenyl indoles (20), a large number of polycyclic aromatic hydrocarbons (21), and benzopyrenediol epoxide (22) also intercalate into the DNA.

The assay architecture is analogous to a protocol generally used for competitive immunoassay, whereby an intercalating dye competes with an analyte for a binding site on the DNA or is displaced by the analyte. The initial guide for this study was a US patent by Richardson and Schulman (23), who used the classic intercalators acridine orange, ethidium bromide, and proflavin and calf

From: *Methods in Biotechnology, Vol. 7: Affinity Biosensors: Techniques and Protocols*
Edited by: K. R. Rogers and A. Mulchandani © Humana Press Inc., Totowa, NJ

thymus DNA to measure small quantities of the drug actinomycin D. The competitive binding assay described here used DNA–acridine orange as a competitive agent to the intercalating test compound. The fluorescence from unbound acridine orange is not polarized because of the random orientation and free rotation of the dye molecules. For a freely rotating molecule the polarization of the fluorescence will be completely random, even if the excitation light is polarized. After the acridine orange intercalates into the DNA, its orientation is fixed and it is unable to freely rotate. The fluorescence emitted from the bound acridine orange will then have the same or similar polarization as the excitation light. The displacement of intercalated acridine orange by a carcinogen is monitored by a reduction in the polarized fluorescence intensity. One advantage of this method is that any test molecule, fluorescent or not, that binds to DNA is detected by the displacement of the fluorescent intercalator. The action is monitored by using excitation and emission wavelengths specific for the fluorescent intercalator. In most cases, any fluorescence of test compounds will not interfere if they do not strongly overlap with the chosen detector dye; also, the quantum efficiency of the dyes will be orders of magnitude greater, reducing interferences.

Advantages of using fluorescence polarization and DNA intercalation are its rapid analysis time, simple experimental apparatus, and good sensitivity as a result of the signal amplification of the multiple binding sites in the DNA. The ability to measure many varieties of carcinogens allows for use in surveys and in screening environmental sites for carcinogen contamination. Present methods recommended by the Environmental Protection Agency (EPA) for the collection and analysis of airborne carcinogens require a 24-h air sampling time plus analysis time, typically with gas chromatography or mass spectroscopy instruments.

2. Materials

1. The fluorescence polarization measurements are made using an SLM 8000C scanning spectrofluorometer, manufactured by SLM Aminco Instruments, Inc.* (Urbana, IL). Other spectrofluorometers with similar specifications and components can also be used. This instrument uses a 450-W xenon arc lamp as the excitation source and a double-grating monochromator for the selection of the excitation wavelength. A single-grating monochromator monitored the fluorescence. The excitation and emission paths contained adjustable Glan-Thompson polarizers and the normal 90° fluorescence geometry was used. Photomultipliers monitored two channels; a reference channel monitoring the xenon lamp, consist-

---

*Certain commercial equipment, instruments, and materials are identified in this chapter to specify adequately the experimental procedure. In no case does such identification imply recommendation or endorsement by the national Institute of Standards and Technology, nor does it imply that the material or equipment is necessarily the best available for that purpose.

ing of a concentrated solution of rhodamine B, which served as a quantum counter, and the fluorescence signal channel. The fluorescence signal was measured and normalized by the reference signal to minimize the effects of lamp fluctuations.

2. A UV-VIS spectrophotometer, used for measurement of the DNA absorption at 260 nm for concentration determinations, is available from Perkin-Elmer Corp. (Norwalk, CT), Milton Roy (Rochester, NY), and many other instrument companies.

3. Standard buffer: 8 mM Tris-HCl, 50 mM NaCl, and 1 mM EDTA in nanopure distilled water; adjust pH to 7.0 with 3 M HCl. Autoclave the buffer for 20 min at 20 psi (*see* **Note 1**).

4. Calf thymus DNA (sodium salt): Take one vial (2 mg) and dissolve in 50 mL of buffer by placing the DNA and buffer in a screw-capped vial and agitate for at least 12 h (*see* **Note 2**). Determine the concentration of DNA in solution by measuring the absorption at 260 nm in a 1-cm path length quartz cuvet (1.00 absorption units of duplex DNA is assumed equal to 50 μg/mL of DNA in solution) *(25)*. The typical DNA concentration range is between 32 and 33 μg/mL.

5. Dilute suspension of glycogen. *See* **Note 3** for preparation and usage.

6. Acridine orange: 2×10^{-5} M in buffer.

7. Test compounds (*see* **Note 4**): Make stock solutions at reasonably high concentrations. Polycyclic aromatic hydrocarbons are not water soluble and must be dissolved in absolute ethanol (*see* **Note 5**). Concentrations of the carcinogens and noncarcinogens prepared in ethanol for this study were: benzo[j]fluoranthrene, 4.36×10^{-5} M; dibenz[a,h]anthracene, 8.96×10^{-5} M; benzo[a]pyrene, 1.98×10^{-5} M; (NCI chemical carcinogen repository, Kansas City, MO); and naphthalene, 7.81×10^{-3} M; anthracene, 5.51×10^{-3} M, and 1,2,3,4,5,6,7,8-octahydronaphthalene, 1.68×10^{-5} M (Aldrich, Milwaukee, WI). *See* **Note 6** for storage requirements.

3. Methods

3.1. Spectrofluorometer Polarization Calibration

Both excitation and emission polarizers could be adjusted to transmit either vertically (0°) or horizontally (90°) polarized light. Both excitation and emission monochromator gratings have different transmission efficiencies when interacting with vertically and horizontally polarized light. As a result, the excitation monochromator partially polarizes the excitation light beam. Thus, rotation of the excitation polarizer to the horizontal (H) or vertical (V) positions yields different intensities of the excitation beam. Likewise, rotation of the emission polarizer changes the effective response of the emission detector. The result is that the measured signals are not the actual values of the parallel (I‖) and perpendicular (I⊥) intensities needed for the polarization calculation.

To calculate the actual intensity ratio (I‖/I⊥) we need to determine the G factor, which is the ratio of sensitivities of the detection system for vertically and horizontally polarized light:

$$G = S_V / S_H \qquad (1)$$

The G factor is dependent on the emission wavelength. Further discussion on polarization measurements and the G factor can be found in the book by Lakowicz *(26)*.

3.1.1. Determination of G

When a dilute suspension of glycogen is used as a scattering sample and excited with polarized excitation light, the scattered light will be 100% polarized. If a high concentration of glycogen is used, multiple scattering will lead to decreased polarization values. The procedure for determining the G value and calculating the polarization is as follows:

1. Fill a fluorescence cuvet with a dilute suspension of glycogen and place into the cell holder of the fluoriometer.
2. Adjust excitation wavelength to 530 nm and emission monochromator to 530 nm (fluorescence maximum for acridine orange) to measure the elastically scattered light.
3. Obtain measurements of the scattered light with different vertical and horizontal polarizer positions on the excitation and emission paths. Data are taken for 30 s with 1 s integration intervals at each polarization setting. Six data sets are obtained with the excitation and emission polarizers in the following positions: (V,V), (V,H), (V,V), (H,V), (H,H) and (V,V) as shown in **Fig. 1**. Obtain the average of the center 25 s (25 points), as indicated by the arrows in **Fig. 1** (*see* **Note 7**), to obtain a value for each signal $I_{EX,EM}$ (excitation polarization, emission polarization). The G value (system polarization response) is calculated by:

$$\frac{I_{HV}}{I_{HH}} = \frac{S_V}{S_H} = G \qquad (2)$$

The G value is then used to obtain the actual values of the parallel ($I\parallel$) and perpendicular ($I\perp$) intensities, unbiased by the detection system, for the calculation of the polarization. The G value should be determined every day.

3.1.2. Determination of Polarization

1. To determine the polarization after obtaining the G value, the data using vertically polarized excitation light, I_{VV} and I_{VH}, obtained in the previous section, are used with the G value to obtain the corrected (for instrument response) parallel and perpendicular intensities:

$$\frac{I_{VV}}{I_{VH}} \frac{1}{G} = \frac{I\parallel}{I\perp} \qquad (3)$$

2. Then use the corrected intensities ($I\parallel/I\perp$) to calculate the anisotropy:

$$r = \frac{(I\parallel/I\perp) - 1}{(I\parallel/I\perp) + 2} \qquad (4)$$

3. The anisotropy is used to calculate the polarization:

$$p = \frac{3r}{2 + r} \qquad (5)$$

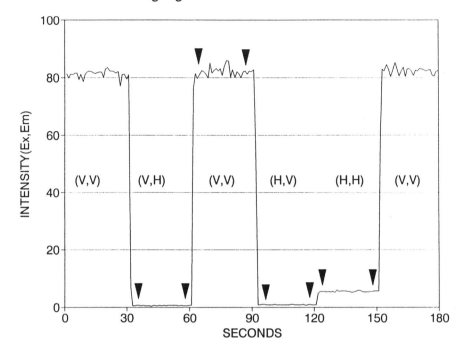

Fig. 1. Scattering signals from glycogen at 530 nm at different polarizer settings for calculation of the *G* value. To calculate *G* the values for (H,V) and (H,H) are used and for the polarization the values for (V,V) and (V,H). Arrows indicate times at which signal was averaged.

The measured value of the polarization for the glycogen sample should be 0.98 or larger (*see* **Note 8**).

3.2. Calibration Curves

All the measurements are made in a 1-cm path length quartz cuvet in the spectrofluorometer. The calibration curve is generated by measuring the decrease in the polarization of acridine orange as a test compound is added to the DNA–acridine orange complex. The test compound intercalates into the DNA and displaces the acridine orange, thereby reducing its polarization. This decrease is proportional to the test compound concentration.

1. Make a stock solution of the test molecule in ethanol or other suitable solvent (*see* **Note 5**) at a relatively high concentration ($\approx 10^{-3} - 10^{-2}$ *M*); the exact value will be dependent on solubility.
2. Make serial dilutions of stock solution in buffer containing 5% solvent down to approx 1×10^{-10} *M* (*see* **Note 9**).
3. Set the excitation monochromator to the excitation wavelength of acridine orange, 490 nm.

4. Place cuvet into cell holder containing magnetic stir bar and add 1.8 mL buffer and 50 μL of 2×10^{-5} M acridine orange to cuvet with constant stirring.

5. Using vertically polarized excitation light at 490 nm measure the fluorescence emission at 530 nm with the emission polarizer set at 0°(V) and then 90°(H). Calculate the polarization using the previously determined G value. This is the value for the free acridine orange polarization.

6. Add 1.6 μg of calf thymus DNA to the cuvet while stirring. Wait 5 min, then measure polarization as in **step 5**. This is the polarization of the intercalated acridine orange.

7. Add 200 μL of the most dilute molecule solution (1×10^{-10} M), wait 5 min, then measure polarization. If no decrease in polarization is observed, continue by adding 200 μL of each higher concentration until a polarization change is observed.

8. Note the concentration at which a change is observed. Prepare a new cuvet containing buffer, DNA, and acridine orange and remeasure its polarization.

9. Starting with the test solution having a concentration an order of magnitude lower than that at which the polarization first decreased, make a series of increasing additions, starting with 5 μL. As the volumes increase switch to higher concentration solutions to minimize volume added. Record all volumes and concentrations added.

10. Calculate the molar concentration of test molecule in the cuvet from the volumes and concentrations added for each polarization point. The analytical curve generated by these measurements is shown in **Fig. 2**.

3.3. Examples

For any molecule studied the above steps are the same; differences occur only in the volumes and concentrations added for a specific test molecule. The general shape of the calibration curve will remain the same; however, the detection limit will be test molecule-dependent.

In **Fig. 3**, the analytical plots for three noncarcinogens and three carcinogens, there are two distinct regions for the curves. The first is where the polarization stays relatively constant with increasing concentration, and in the second region the polarization drops linearly with increasing concentration. The point at which the polarization starts to decrease is the limit of detection (LOD) for a given molecule. There is a great difference in the LODs of these molecules, ranging from 3.7×10^{-5} M for naphthalene to 4.7×10^{-8} M for benzo[a]pyrene. These could also be considered to be a measure of the relative affinities of the molecules for the displacement of acridine orange from the DNA. Using these calibration curves the range of concentrations that could be measured would be between 4.7×10^{-8} and 1×10^{-6} M for benzo[a]pyrene and 3.7 $\times 10^{-5}$ and 3×10^{-4} M for naphthalene.

This technique was used to generate curves and detection limits for the other molecules examined in this paper: anthracene, 2.6×10^{-6} M, 1,2,3,4,5,6,7,8-octahydronaphthalene, 1.2×10^{-6} M; dibenz[a,h]anthracene, 1.8×10^{-7} M; and

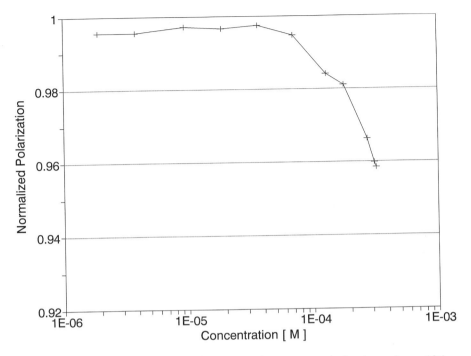

Fig. 2. Typical analytical curve of concentration versus polarization using calf thymus DNA and acridine orange.

benzo[*j*]fluoranthrene, 5.3×10^{-8} *M* as shown in **Fig. 3**. As seen in **Fig. 3**, the response to analytes covers several orders of magnitude of concentration, and a wide range of detection limits is apparent. The small sample size required for these measurements yield excellent absolute sensitivities. For two highly carcinogenic molecules, benzo[*a*]pyrene and dibenz[*a,h*]anthracene, the LODs were found to be 6 and 25 ppb, respectively. These two compounds have been listed as priority pollutants along with 14 other polynuclear aromatic hydrocarbons (PAHs). The typical urban background concentration of carcinogenic PAHs in soils is on the order of 1000–3000 µg/kg (1–3 ppm) *(26)*. In another study using supercritical fluid extraction and gas chromatography-mass spectrometry, the concentration of benzo[*a*]pyrene at 24 different sites was found to range from 9 to 7600 ng/g *(27)*. The highest ambient concentrations of carcinogenic PAHs in soils have been reported for road dust, which can contain levels of 8–336 ppm *(26)*. This study also estimated the concentration of carcinogenic PAHs in air to range from 3.5 ng/m^3 in rural areas to 50 ng/m^3 in highly urbanized areas. The limits of detection for these two carcinogens obtained with this method have the sensitivity required for environmental measurements with the simplicity needed for a field survey instrument.

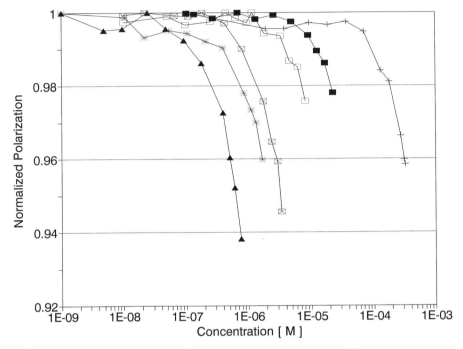

Fig. 3. Analytical curves with calf thymus DNA and acridine orange for: +, Naphthalene; ■, Anthracene; □, 1,2,3,4,5,6,7,8-octahydronaphthalene; ⊠, Dibenz-[*a,h*]anthracene; *, Benzo[*j*]fluoranthrene, ▲, Benzo[a]pyrene.

3.4. Conclusions

We have demonstrated that polycyclic aromatic hydrocarbons can be rapidly measured using the DNA intercalation—fluorescence polarization technique. The experimental apparatus required is relatively inexpensive and the procedures are simple. The use of DNA yields a simple, highly sensitive detector because of the large number of binding sites filled with strongly fluorescence dye. The broad range of molecules that can be measured indicate that this technique can be used for rapid screening of sites of environmental contamination. For air pollution studies, the collection filters can be extracted with ethanol, then directly diluted with buffer, eliminating time-consuming methods required for other analytical techniques.

4. Notes

1. After cooling, the autoclaved buffer and all solutions made using the buffer in **Subheading 2., item 3**, should be refrigerated at 5°C.
2. The DNA is completely dissolved by placing the vial on a Nutator mixer (VWR Scientific) overnight (*see* **Subheading 2., item 4**).

3. The glycogen used for a polarization scattering standard is an animal liver starch, a high-mol-wt polymer. The glycogen obtained from Sigma is a white granular powder. To make a stock scattering solution a few particles of glycogen should be placed in 250 mL of distilled water in a screw-cap bottle and shaken to dissolve. The solution should be allowed to settle for 20 min before use. A cuvet of scattering solution should be used once and discarded and the remainder of the stock stored under refrigeration. After filling the cuvet, the glycogen should warm up to room temperature before determining G (*see* **Subheading 2., item 5**).

4. Many of the molecules used in **Subheading 2., item 7** are carcinogenic or mutagenic. Proper precautions should be used when handling and when disposing of these materials.

5. Other solvents, such as methanol, acetone, and so forth, may be suitable for dissolving test molecules. Specific solubility data on individual molecules should be obtained.

6. Polycyclic aromatic hydrocarbons are sensitive to light and oxygen and should be stored in opaque bottles under nitrogen (*see* **Subheading 2., item 7**).

7. The center 25 s was used to ensure only one polarization was observed. This eliminated data gathered during rotation of the polarizers (*see* **Subheading 3.1.1., step 3**).

8. A value smaller than 0.98 is caused by multiple scattering occurring at high glycogen concentration. If this occurs the glycogen suspension should be diluted and scattering measurements should be repeated until a polarization of at least 0.98 is obtained. At this point, where the polarization ≥ 0.98, G can be accurately determined (*see* **Subheading 3.1.2., step 3**).

9. Making serial dialations helps prevent the test molecule from precipitating out in the buffer and is also required to prevent the denaturing of DNA by the pure solvent. Solvents should be examined for DNA stability (*see* **Subheading 3.2., step 2**).

Acknowledgments

The author thanks the Environmental Protection Agency, Interagency agreement #DW13937298-01-0, for their support of this work, and Manana Gueguetchkeri for preparation of the figures and helpful discussions.

References

1. Michaelis, L. (1947) The nature of the interaction of nucleic acids and nuclei with basic dyestuffs. *Cold Spring Harb. Symp. Quant. Biol.* **12,** 131–142.
2. Peacock, R. A. and Skerrett, J. N. H. (1956) The interaction of aminoacridines with nucleic acids. *Trans. Faraday Soc.* **52,** 261–279.
3. Bresloff, J. L. and Crothers, D. M. (1981) Equilibrium studies of ethidium-polynucleotide interactions. *Biochemistry* **20,** 3547–3553.
4. Scatchard, G. (1949) The attractions of proteins for small molecules and ions. *Ann. NY Acad. Sci.* **51,** 600–672.
5. Cavalier, L. F., Rosoff, M., and Rosenberg, B. H. (1956) Studies on the structure of nucleic acids. X. On the mechanism of denaturation. *J. Amer. Chem. Soc.* **78,** 5239–5247.

6. Wilson, W. D. and Jones, R. L. (1982) Intercalation in biological systems, in *Intercalation Chemistry* (Whittingham, M. S. and Jacobson, A. J., eds.), Academic, New York, pp. 445–501.

7. Richardson, C. L. and Schulman, G. E. (1981) Competitive binding studies of compounds that interact with DNA utilizing fluorescence polarization. *Biochim. Biophys. Acta.* **652,** 55–63.

8. Shahbaz, M., Harvey, R. G., Prakash, A. S., Boal, T. R., Zegar, I. S., and LeBreton, P. R. (1983) Fluorescence and photoelectron studies of the intercalative binding of benz[*a*]anthracene metabolite models to DNA. *Biochem. Biophys. Res. Comm.* **112,** 1–7.

9. Zegar, I. S., Prakash, A. S., and LeBreton, P. R. (1984) Intercalative DNA binding of model compounds derived from metabolites of 7,12-dimethylbenz[*a*]anthracene. *J. Biomol. Struct. Dyn.* **2,** 531–542.

10. LeBreton, P. R. (1985) The intercalation of benzo[*a*]pyrene and 7,12- dimethylbenz[*a*]anthracene metabolites and metabolic model compounds into DNA, in *Polycyclic Hydrocarbons and Carcinogenesis*, Symposium Series 283, American Chemical Society, Washington, DC, 209–238.

11. Dinesen, J., Jacobson, J. P., Hansen, F. P., Pedersen, E. B., and Eggert, H. (1990) DNA intercalating properties of tetrahydro-9-aminoacridines. Synthesis and [23] Na NMR spin-lattice relaxation time measurements. *J. Med. Chem.* **33,** 93–97.

12. Nordmeier, E. J. (1992) Ethidium bromide binding to calf thymus DNA: implications for outside binding and intercalation. *J. Phys. Chem.* **96,** 6045–6055.

13. Neidle, N., Pearl, L. H., Herzyk, P., and Berman, H. M. (1989) A molecular model for proflavine-DNA intercalation. *Nucleic Acids Res.* **16,** 8999–9016.

14. Zimmerman, S. C., Lamberson, C. R., Cory, M., and Fairley, T. A. (1989) Topologically constrained bifunctional intercalators: DNA intercalation by a macrocyclic bisacridine. *J. Amer. Chem. Soc.* **111,** 6805–6809.

15. Tanious, F. A., Veal, J. M., Buczak, H., Ratmeyer, L. S., and Wilson, W. D. (1992) DAPI (4',6-Diamidion-2-phenylindole) binds differently to DNA and RNA: minor-groove binding at AT sites and intercalation at AU sites. *Biochemistry* **31,** 3103–3112.

16. Lerman, L. S. (1962) The structure of the DNA-acridine complex. *Proc. Natl. Acad. Sci. USA* **49,** 94–101.

17. Lerman, L. S. (1964) Acridine mutagens and DNA structure *J. Cell. Comp. Physiol.* **64(Suppl. 1),** 1–18.

18. Kapuscinski, J. and Darzynkiewics, Z. (1987) Interactions of acridine orange with double stranded nucleic acids. Spectral and affinity studies. *J. Biomol. Struct. Dyn.* **5,** 127–143.

19. McFadyen, W. D., Sotirellis, N., Denny, W. A., and Waklin, L. P. G. (1990) The interaction of substituted and rigidly linked diquinolines with DNA. *Biochem. Biophys. Acta* **1048,** 50–58.

20. Wilson, W. D., Tanious, F. A., Barton, H. J., Strekowski, L., Boykin, D. W., and Jones, R. L. (1989) Binding of 4',6-diamino-2-phenylindole (DAPI) to GC and mixed sequences in DNA: intercalation of a classical groove-binding molecule. *J. Amer. Chem. Soc.* **111,** 5008–5010.

21. Harvey, R. G. and Geacintov, N. E. (1988) Intercalation and binding of carcinogenic hydrocarbon metabolites to nucleic acids. *Acc. Chem. Res.* **21,** 66–73.
22. Kim, S. K., Geacintov, N. E., Brenner, H. C., and Harvey, R. G. (1989) Identification of conformationally different binding sites in benzo[*a*]pyrene diol epoxide-DNA adducts by low-temperature fluorescence spectroscopy. *Carcinogenisis* **10,** 1333–1335.
23. Richardson, C. L. and Schulman, G. E. (1981) Intercalation inhibition assay for compounds that interact with DNA or RNA, United States patent #4,257,774; March 24.
24. Gibco BRL (1991) Catalogue and Reference Guide, Life Technologies, Inc., Gaithersburg, MD.
25. Lakowicz, J. R. (1984) *Principles of Fluorescence Spectroscopy.* Plenum, New York, pp. 111–131.
26. Menzie, C. A., Potocki, B. B., and Santodonato, J. (1992) Exposure to carcinogenic PAH's in the environment. *Environ. Sci. Technol.* **26,** 1278–1284.
27. Yang, Y. and Baumann, W. (1995) Seasonal and areal variations of polycyclic aromatic hydrocarbon concentrations in street dust determined by supercritical fluid extraction and gas chromatography-mass spectrometry. *Analyst* **120,** 243–248.

12

ISFET Affinity Sensor

Geert A. J. Besselink and Piet Bergveld

1. Introduction
1.1. General Introduction

The so-called ion-step method represents a newly developed measurement concept for potentiometric detection and quantification of adsorbed biomolecules in which modified ion-sensitive field-effect transistors (ISFETs) are used. The ion-step method is based on a dynamic measuring principle, whereas many other potentiometric methods are static and measure in a state of thermodynamic equilibrium. A number of authors report on measuring protein adsorption by equilibrium potentiometry, but in all measurements the observed responses were poor *(1,2)*. The new measuring method is therefore an important alternative method for detection of adsorbed protein. In **Subheading 1.2.**, a short explanation will be given for the disappointing results obtained with detection of protein by equilibrium potentiometry. Thereafter, the ion-step method will be described briefly and at the end of this introduction, the use of an ISFET affinity sensor for the measurement of heparin will be described.

1.2. Restrictions of Equilibrium Potentiometry for Protein Detection

Attempts to detect proteins by using ISFETs (and other potentiometric ion sensors) were not very successful, as appears from the literature *(1,2)*. To explain these negative results, distinction must be made between ion sensors to which protein is directly adsorbed, and ion sensors that support a protein-containing membrane.

Studies concerning protein adsorption to bare devices, such as ISFETs, were started from the expectation that proteins should modulate the static ISFET response, considering the fact that proteins carry electrical charge. The assump-

From: *Methods in Biotechnology, Vol. 7: Affintiy Biosensors: Techniques and Protocols*
Edited by: K. R. Rogers and A. Mulchandani © Humana Press Inc., Totowa, NJ

tion was made that the inherent charge of protein molecules, when attached to the gate of an ISFET, would create an external field that would be sensed by the FET structure. Recently, papers appeared in which it was explained and proven that such an operational mechanism cannot exist because counter ions shield the charged protein molecules, thus resulting in an absence of an external electric field beyond a distance determined by the Debye length of the sample solution *(3)*. Only changes in charge amount that occur within the order of a Debye length of the ISFET surface can be detected. The Debye length, defined as the distance at which the electrostatic field has dropped to $1/e$ of its initial value, is strongly dependent on the ionic strength of the solution (**Eq. 1**).

$$L_D = \frac{0.304}{\sqrt{c^e}} \tag{1}$$

where L_D is the Debye length (nm), and c^e is the salt concentration (1:1 salt, *M)*. In a physiologic salt solution the Debye length is limited to about 0.8 nm, which is a small value as compared with the dimensions of protein molecules (about 10 nm). It was therefore concluded that, from a theoretical point of view, it is very difficult to detect the presence of a directly adsorbed layer of proteins on an ISFET in a situation of thermodynamic equilibrium.

A membrane phase with immobilized protein molecules contains a certain amount of fixed charge that stems from ionized amino acid side-chain groups. The presence of the membrane-fixed charge, together with the condition of electrochemical equilibrium, gives rise to a Donnan equilibrium between the membrane and the adjacent electrolyte, with as one of its aspects a potential difference between both, called the Donnan potential. Measuring this Donnan potential may be used as a method to detect membrane-bound proteins. However, an ion sensor that is Nernstian sensitive for its potential determining ion (pdi) (i.e., $\alpha = 1$ in **Eq. 2**) cannot measure the Donnan potential. The measured signal of a sub-Nernstian-sensitive ion sensor ($0 < \alpha < 1$) is partly determined by the Donnan potential and partly by the activity of its pdi in the adjacent phase (**Eq. 2**), which means that with such a sensor, equilibrium potentiometry may be a useful method for detection of adsorbed proteins. The α value of ISFET sensors with different gate oxides varies from 0.05 (SiO_2, at pH 2.0) to 0.99 (Ta_2O_5).

$$\Phi^s - \Phi^e = (1 - \alpha) \cdot (\Phi^m - \Phi^e) + \alpha \frac{RT}{z_{pdi}F} \ln a^e_{pdi} + K \tag{2}$$

where $\Phi^s - \Phi^e$ is the ion-sensor surface potential minus the electrolyte bulk potential (V), α is the proportionality factor related to the sensitivity of the sensor for its pdi, $\Phi^m - \Phi^e$ is the Donnan potential (V), z_{pdi} is the valence of the pdi, a^e_{pdi} is the activity of the pdi in electrolyte bulk (mol/m³), and K is a constant. The other symbols have their usual meaning.

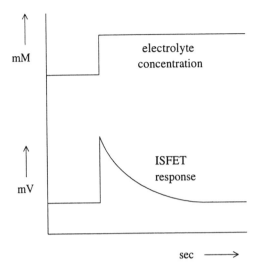

Fig. 1. Schematic representation of the stepwise change of salt concentration and the resulting transient change of the ISFET potential as determined with the ion-step measurement setup.

However, several conditions must be fulfilled for this to happen *(4)*:

1. No electrolyte solution is allowed to be present between the membrane and sensor, forming an interlayer.
2. The membrane must be of sufficient thickness to build up a Donnan potential.
3. The ratio between fixed-charge concentration and the salt concentration in the electrolyte bulk must be sufficiently large.

It can be concluded that detection of protein by equilibrium potentiometry is not a very simple, straightforward approach.

1.3. The Ion-Step Method

The ion-step measuring method to determine the concentration of proteins (or more generally of charged molecules) was first introduced by Schasfoort et al. *(5)* and was further developed by Van Kerkhof et al. *(6)* and Eijkel *(4)*. To execute the ion step, the ISFET is placed in a flowthrough system, and the electrolyte solution flowing over the ISFET gate is suddenly changed to a solution having a higher ion concentration than the initial one but with the same pH. Application of this ion step appeared to induce a transient potential change, as measured by the ISFET (**Fig. 1**). The amplitude and the time of decay of this transient potential contain information concerning the fixed charge concentration of the ISFET surface or the concentration of acidic and basic groups in the ISFET-connected membrane structure. In practice, only the amplitude is used.

Advantage can be taken of the small planar performance of the ISFET and the fast response to local pH changes. In general all affinity-based reactions with molecules having ionizable surface groups can be monitored using the ISFET stimulus-response system. In addition, the concentration of neutral molecules with a charged label can be measured.

To elucidate the general mechanism of the ISFET response very shortly, three types of ISFET modifications will be distinguished as follows:

1. Directly immobilized, small (bio)molecules.
2. Indirectly immobilized proteins, which are bound to a membrane, deposited on the ISFET.
3. Directly immobilized (monolayer of) proteins.
   a. Direct immobilization of small charged molecules, such as protamine (mol-wt 4 kDa), can be regarded as a surface modification that changes the electrical potential at the sensor-solution interface. A changed surface potential is reflected in a change of the ISFET response amplitude when ion stepping is performed, and the corresponding mechanism lies in a salt-induced increase of the double-layer capacitance, as clarified by Van Kerkhof et al. *(7)*.
   b. After application of an ion step, first the Donnan potential at the electrolyte-membrane interface changes and consequent to the changing Donnan potential all ions, including the proton, are redistributed between electrolyte and membrane according to the new Donnan equilibrium. After proton re-equilibration, the ISFET potential has returned to its original value. Proton re-equilibration is delayed by the release or uptake of protons from protonable groups in the membrane, in particular of adsorbed protein molecules. If re-equilibration takes a sufficiently long time (>>1 s) when compared with the establishment of the Donnan potential (<1 s), the ISFET response amplitude can be used as a measure of the concentration of immobilized (protein-derived) charge (*see* **Eq. 3**). The membrane type that has been used mostly in our group in combination with ISFET protein sensor work is a microporous polystyrene-agarose membrane that consists of polystyrene beads (diameter size of 0.11 μm) that are embedded in an agarose matrix.

$$\Delta\Phi^s_{extr} = \frac{RT}{F} \ln\left(\frac{r_{D,h}}{r_{D,l}}\right) \tag{3}$$

$$r_D = \frac{\sqrt{4(c^e)2 + (zc^m_{fixed})2} + zc^m_{fixed}}{2c^e} \tag{3a}$$

where $\Delta\Phi^s_{extr}$ is the maximal ISFET response amplitude (V), r_D is the Donnan ratio (subscripts l and h refer to the low and high salt concentration), c^e is the salt concentration in bulk electrolyte (*M*), and zc^m_{fixed} is the net fixed charge concentration including its valence (mol/m³).

   c. Immobilization of protein molecules directly to the ISFET surface may lead to modification of the surface potential of the ISFET (*see* **point 1**). Because the dimensions of proteins (up to 10 nm) outsize the thickness of the double layer also, a Donnan potential may be formed in the case of immobilized, net charged protein molecules. Therefore, the ion-step response of an ISFET with an adsorbed monolayer of protein will be the combined result of the processes indicated under (1) and (2). However, it must be noted that a protein layer for which the thickness does not exceed a monolayer is too thin for sustaining a full buildup of a Donnan potential.

1.4. ISFET Affinity Sensor

As a first application of the ion-step measuring method, an ISFET-based heparin sensor was developed *(7,8)*. Heparin is a highly negatively charged polysaccharide that is used clinically to delay clotting of blood. The relationship between the dosing of heparin and the resulting biological activity is poorly understood and differs between individual patients. Therefore, the heparin treatment must be carefully monitored. Currently this is a cumbersome and time-consuming procedure using laboratory analysis equipment. It would be useful to replace this method by a simpler one that can be carried out by nursing personnel. For this purpose, a heparin sensor has been developed based on the ion-step method.

Protamine, immobilized directly onto the ISFET surface, was used as the ligand for the binding of heparin. Protamine is a highly positively charged peptide (mol wt 4000 kDa) that is used in clinical practice to neutralize heparin already present in the blood circulation. The interaction between heparin and protamine is an electrostatic interaction: the negatively charged heparin binds to the positively charged protamine. When a protamine-coated ISFET is immersed in a blood sample containing heparin, mainly heparin binds to the ISFET surface. If a fixed incubation time is used, then the amount of bound heparin is a measure of its concentration in the blood sample. Experiments show that a certain amount of nonspecific binding of other components from the blood plasma occurs to the modified ISFET surface. It is of course of importance to keep this nonspecific binding to a minimum. In **Fig. 2** three typical responses on an ion-step are shown. Curve 1 is the ion-step response of a bare ISFET, curve 2 represents the response of an ISFET with a layer of protamine, and curve 3 is the response of the same ISFET, but after incubation in buffer to which a certain amount of heparin was added. The difference between the amplitude of curves 2 and 3 (ΔA) is a measure for the amount of bound heparin. After each measurement the sensor can be regenerated by rinsing the ISFET in 4 M NaCl. Then all bound heparin is removed from the ISFET surface, including the aspecifically bound molecules, without removing the protamine. In **Fig. 3**, the variable ΔA is plotted as a function of the heparin concentration, measured in plasma samples with added quantities of heparin. As is clear from this figure, a linear relation exists between heparin concentration and ΔA, whereas the substantial ΔA at 0 U heparin/mL corresponds to

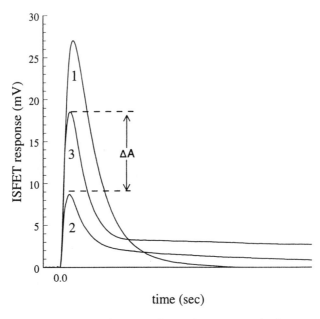

Fig. 2. ISFET responses on an ion-step. Curve 1 represents the ion-step response of a bare ISFET, curve 2 is the response of an ISFET with a monolayer of protamine, and curve 3 gives the response of the same ISFET after incubation in buffer containing 0.9 U/mL heparin.

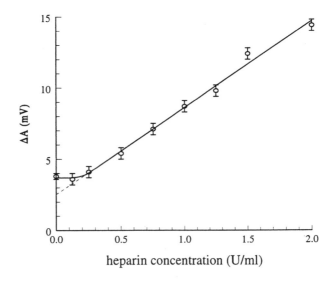

Fig. 3. The change in amplitude of the ion-step response (ΔA) before and after incubation in human plasma, containing heparin, as a function of the heparin concentration.

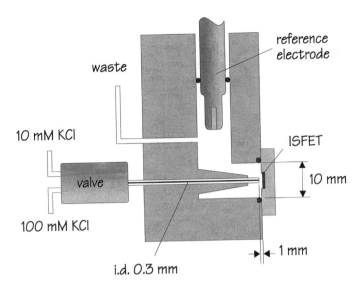

Fig. 4. Cross-section (in side-view) of the wall-jet cell with mounted ISFET device.

nonspecific binding of plasma components other than heparin. This nonspecific binding to the protamine-coated surface reduces the detection limit of heparin in plasma to about 0.25 U/mL. An activity of 0.25 U/mL corresponds to an absolute heparin concentration of about $13.3 \times 10^{-8}\ M$, assuming an average molecular weight of 15 kDa and a value of 8 µg per unit activity. The heparin-concentration range that can be detected in plasma, when using protamine-coated ISFETs, lies between 0.25 and 2 U/mL, with an accuracy of 0.08 U/mL. Therefore, the protamine-covered ISFET sensor can be used for heparin monitoring during treatment of postoperative thrombosis and embolism (which treatments have a therapeutic concentration range of 0.2–0.7 U/mL).

2. Materials
2.1. Measurement Set-Up

1. Wall-jet cell: Plexiglas housing with 0.3-mm id feed tubing and a relatively wide outlet chamber (*see* **Note 1**), rubber O-ring, three-way solenoid control valve (LFYA1201032H; Lee, Westbrook, CT), and specifications as indicated in **Fig. 4**.
2. Glass bottles (0.5 or 1 L) with three-way valve caps (Omnifit Limited, Cambridge, UK).
3. Technical nitrogen.
4. Analog pressure regulator (range up to 15 psi; Omnifit Limited).
5. PTFE tubing (0.8-mm id, 1.6-mm od).
6. Flowmeter (Brooks Shorate Purgemeter GT 1355, Brooks Instrument BV, Veenendaal, The Netherlands).
7. Ag/AgCl reference electrode (type REF200; Radiometer, Copenhagen, Denmark).

8. Laboratory ISFET amplifier of the source-drain follower system (Electro Medical Instruments, Enschede, The Netherlands).
9. Nicolet 310 memory oscilloscope (Nicolet, Madison, WI).
10. Vu-Point software package (Maxwell Laboratories Inc., La Jolla, CA).

2.2. Measurement Devices

1. Wafer containing conventional SiO_2 ISFETs.
2. Tantalum; evaporation slug, 0.6-cm diameter × 0.6-cm length.
3. HF/HNO_3 etching mixture; consisting of 1 part 50% HF, 1 part 69% HNO_3, and 4 parts demineralized water.
4. 3-Aminopropyl triethoxysilane (APS): a 0.5% solution in methanol/water (19:1 v/v). This solution was stored at 4°C for at least 1 night and a maximal 3 wk before being used.
5. Polyimide solution: obtained by mixing 1 mL polyimide (Dupont [Bristol, UK] PI2555) with 2 mL N-methylpyrrolidone.
6. Printed circuit board.
7. Hysol epoxy (Dexter, Garching, Germany): mixture of 2 g prepolymer (C8W795) and 0.5 g hardener (H-W796).
8. 0.25% agarose (purified fraction: M_r zero; Bio-Rad, Hercules, CA) in demineralized water.
9. Protamine sulfate (grade X, from salmon); solutions of 0.1 and 10 mg/mL in phosphate-buffered solution (PBS: 140 mM NaCl, 10 mM sodium phosphate, pH 7.4).
10. Polystyrene beads (112 nm; Polysciences, Warrington, PA): supplied as a 2.5% suspension in demineralized water.

2.3. Measurement Protocol

1. Phosphate-buffered saline (PBS): 140 mM NaCl, 10 mM sodium phosphate, pH 7.4.
2. Heparin: Thromboliquine® batch solution (5000 U/mL), diluted with PBS to get solutions with activities up to 100 U/mL.
3. Citrated normal human plasma.
4. Low ion buffer: 10 mM KCl, 0.2 mM HEPES, pH 7.4.
5. High ion buffer: 100 mM KCl, 0.2 mM HEPES, pH 7.4.
6. 4 M NaCl in water.

3. Methods

3.1. Measurement Setup

The measurement setup for the ion-step measurements consists of a computer-controlled flowthrough system. The flow is generated by an effective pressure of nitrogen in the two bottles containing the ion-step solutions in which overpressure is maintained at 0.4 bar by means of an analog pressure regulator that is placed in-line with the bottles. The flowthrough system is equipped with a flowmeter to check the buffer flow during the experiment, and this unit can also be used to control the flow (range: 0.2–6 mL/min) (*see* **Note 2**). The ISFET is mounted in a wall-jet cell (**Fig. 4**) in which the liquid flow is perpendicular to the ISFET surface: To obtain this the ISFET is tightly pressed to a Perspex cube (*see* **Note 3**), in which a narrow-feed tubing and a wide outlet chamber (*see* **Note 1**) are accommodated. The two bottlés containing the two ion-step solutions are connected via a solenoid valve to the measurement cell in such a way that the electrolyte concentration at the ISFET surface can be increased with a rise time (to 90% of the final value) of <200 ms. The distance between the ISFET gate and the valve is about 30 mm. An Ag/AgCl reference electrode, placed downstream, is used to define the potential of the solution. The ISFETs are connected to a source-drain follower, which measures a potential that is proportional to the surface potential of the ISFET (*see* **Notes 4** and **5**), and the output of this amplifier is connected to a Nicolet 310 digital oscilloscope, which has the ability to store recorded curves on a floppy disk. The data can then be analyzed on a PC using the software package Vu-Point.

3.2. Measurement Devices

3.2.1. Fabrication of Encapsulated ISFETs

Wafers with conventional SiO_2 gate oxide ISFETs were fabricated in the cleanroom laboratory of the MESA Research Institute following the usual ISFET processing steps. In our experiments Ta_2O_5 is used as the gate oxide (*see* **Note 6**); this layer is obtained as follows. Tantalum is evaporated on the SiO_2 ISFETs (on wafer level) by means of an E-beam evaporation equipment (from Varian), resulting in a thin (50-nm) tantalum film. Then, photolithography is performed with a "tantalum"-mask and the tantalum that is not covered with photoresist is etched away with the HF/HNO_3 etching mixture. After removal of the photoresist, the tantalum film is oxidized in ambient oxygen (2 h at 600°C), resulting in the required Ta_2O_5 gate oxide. The sensors are covered by a layer of polyimide, except for the sensing and contact area. To improve the adhesion of the polyimide to the Ta_2O_5 surface, aminopropyltriethoxysilane (APS) is used as primer. The polyimide is patterned by a photolithographic process. After dicing the wafer into single chips, the ISFETs are glued to a printed circuit board and the contact pads on the chip are wire-bonded to the metal strips on the support. For encapsulation, the supported ISFET is placed in a Teflon mold and a Teflon screw is placed on top of the gate (*see* **Note 7**). The mold is filled with Hysol, an epoxy resin. The epoxy is cured at 60°C for at least 1 h, after which the screw and the mold are removed. Around the gate of the encapsulated ISFET, a circular area with a diameter of 2.5 mm and a depth of about 150 μm is left uncovered in the encapsulation procedure. The resulting Ta_2O_5 ISFETs show a response of about −58 to −59 mV/pH.

3.2.2. Immobilization of Protamine

Protamine is immobilized to the ISFETs by physical adsorption, using either method (a) or (b):

(a) Direct binding to the gate oxide surface:

1. ISFETs are immersed in a solution of 10 mg/mL protamine sulfate in PBS for 16 h (*see* **Note 8**).
2. The ISFETs are rinsed in 4 M NaCl and subsequently stored in PBS at 4°C.

(b) Binding to polystyrene beads that are layered over the ISFET gate:

1. A suspension of polystyrene beads (2.5% w/v) is mixed with a 0.25% agarose solution (1:1) at 40–50°C.
2. Portions of 3 µL polystyrene-agarose suspension are casted on top of the gate area of the ISFETs (*see* **Note 9**).
3. The devices are maintained at 4°C for 12–16 h (to allow slow evaporation of the water).
4. Subsequently, the device is heated at 55°C (for 1 h). In this way membranes with a thickness of about 10–15 µm (in dry condition) are obtained.
5. Protamine is immobilized in the membrane by physical adsorption by exposing the membranes to a solution of 0.1 mg/mL protamine sulfate in PBS at pH 7.4 for about 16 h.
6. The ISFETs are rinsed in demineralized water and subsequently stored in PBS at 4°C.

3.3. Measurement Protocol

1. The ISFETs with immobilized protamine are mounted in the wall-jet cell of the measurement setup and an ion-step of 10–100 mM KCl is applied (at a buffer flow of 4 mL/min) (*see* **Notes 10** and **11**). The ion-stepping solutions are buffered with 0.2 mM HEPES at pH 7.4 (*see* **Notes 12–14**). Each time the ion-step response has to be determined, three to five responses are successively recorded, and the mean amplitude of these responses is calculated.
2. For the determination of heparin concentrations in PBS solutions, 15-mL vessels are used in which the ISFET is placed during the respective incubation time while the solution is not stirred. For the determination of heparin concentrations in plasma, a test tube containing 2 mL normal citrated human plasma is used to which small amounts of a 100 U/mL heparin solution are added to obtain the different concentrations. The ISFET is incubated in heparin-containing buffer or plasma at a fixed incubation time (2 min for the protamine-coated ISFETs; 15 min for the membrane-covered ISFETs).
3. After incubation of ISFETs with heparin, the ISFETs are rinsed in PBS. The ion-step response is recorded and the change in amplitude with respect to the response before incubation with heparin is taken as a parameter.
4. The ISFETs with a protamine monolayer can be regenerated after each measurement by rinsing for about 1 min in 4 M NaCl followed by an equilibration period in PBS (for at least 5 min).

4. Notes
4.1. Measurement Setup

1. The outlet of the wall-jet cell must be relatively wide to minimize the electrical resistance between the ISFET surface and the reference electrode, which is placed downstream of the ISFET (*see* **Subheadings 2.1.** and **3.1.**). When, unfortunately, 50-Hz noise appears to be persistent, the curves can be filtered with a software low-pass filter using a cutoff frequency of 40 Hz for elimination of the 50-Hz main supply interference.

2. Flow regulation can also be realized by using a peristaltic pump at the outlet of the wall-jet cell (which still requires an effective N_2 pressure) instead of using a flowmeter (*see* **Subheadings 2.1.** and **3.1.**). One advantage of the peristaltic pump is its stable flow performance without the need of repeated checking. An advantage of using the flowmeter is that the actual flow performance can be read off immediately and at any time during the experiment.

3. Mounting of the ISFET in the wall-jet cell has to be done in a reproducible and proper manner considering the positioning of the ISFET gate relative to the feed-tube ending (*see* **Subheadings 3.1.** and **3.3.**).

4. When examining slow ISFET responses, for example, in the case of membrane-covered ISFETs, measurements should be carried out in the absence of light. ISFETs are inherently light-sensitive and when light fluctuations occur, the signal may become unstable.

5. No metals should directly contact the solutions; otherwise this can result in electrical leakage currents and hence a voltage response to an ion step.

4.2. Measurement Devices

6. In our group Ta_2O_5 ISFETs are used but we think that other gate oxides, such as Al_2O_3 or Si_3N_4, may also be useful in the ion-step approach. SiO_2 gate oxide ISFETs are not recommended because these show hysteresis and a relatively low pH sensitivity when compared with the other mentioned gate oxides. Furthermore, the pH sensitivity of SiO_2 proves to be strongly pH-dependent whereas that of the other oxides mentioned is very constant in the pH region of 2.0–12.0.

7. Especially during the process of encapsulation (*see* **Subheading 3.2.1.**), the ISFET gate area may become contaminated with epoxy and other compounds. Therefore, care must be taken to use only clean Teflon screws and molds and to prevent contact of the ISFET gate with epoxy (pre)polymer and other volatile compounds. Devices that are contaminated may appear worthless for further use.

8. Recently, we found that incubation of protamine with the bare ISFET at pH 12.0 resulted in a much better adsorption of the peptide when compared with the case when pH 7.4 is used (*see* **Subheading 3.2.2.**). A better coverage of the ISFET surface with protamine may decrease nonspecific binding of, for example, plasma compounds to the resulting surface.

9. Membrane morphology is reported to be more homogeneous when, directly before casting the polystyrene-agarose on the ISFET (*see* **Subheading 3.2.2.**), the former suspension is ultrasonically treated for 2 min.

4.3. Measurement Protocol

10. Flow of the ion-step buffers, applied on the ISFET surface, must be kept constant throughout the measurements (*see* **Subheading 3.3.**). The effect of flow rate on the ion-step response of protamine-coated ISFETs is very profound: At a buffer flow of <3 mL/min, amplitude increases sharply when flow is increased, whereas the amplitude levels off at flow rates >3 mL/min. Therefore, the fixed flow rate should not be chosen to be <3 mL/min.

11. The transient potential amplitude can be increased by increasing the ion-stepping ratio (i.e., the ratio between the salt concentrations of the high and the low ion buffer) (*see* **Eq. 3**).

12. Ion stepping with stronger buffered solutions produces ion-step responses with smaller amplitudes, which compromises the sensitivity of the method. Useful buffer concentrations lie in the range of 0.1–0.5 mM, with buffer species, such as HEPES or Tris.

13. The pH of the low ion buffer and the high ion buffer (*see* **Subheading 3.3.**) must be equal (difference of 0.01 pH unit or less), or else the ISFET potential will not return to its original value, which introduces a pH offset artifact. With small ΔpH differences this may not have consequences regarding the measurement result; however, when the differences are considerable (>0.05 pH unit), the magnitude of the response amplitude may be affected. Proper pH adjustment of the buffer solutions must be done repeatedly during the experiment because the low buffer capacity of the buffer solutions allows pH changes to happen very easily (despite N_2 purging). CO_2 may diffuse through the PTFE tubing wall into the solution, which may cause a local pH change, especially when nonbuffered solutions are used. Residence time of solutions in the tubing should therefore be minimized.

14. The pH of the ion-stepping solutions must be chosen to be sufficiently different from the isoelectric point (pI) of the bioanalyte molecule. The larger the difference between ion-stepping pH and the pI of the bioanalyte, the larger the net charge of the bioanalyte molecules, which augments its detection limit.

15. Air or N_2 bubbles may appear in the tubing and get caught near the ISFET gate, which can cause excessive noise interference. Care should be taken to remove them.

4.4. General Notes

16. Many biosensors are meant as real-time measuring and detection devices. The ISFET-based affinity sensor uses the dynamic principle of an ion-step-induced device signal and, therefore, incubation of the ISFET with sample solutions has to be interrupted at each sampling time to be able to execute the ion step (*see* **Subheading 3.3.**). This approach puts restriction on the application of the ISFET affinity sensor as a real-time measuring device. However, the system is extremely suitable as a one-shot (disposable) sensor, as shown with respect to the heparin sensor.

17. Using a polystyrene-agarose membrane on top of the ISFET enables one to immobilize affinity ligands by the simple approach of passive adsorption. Protein adsorption to the hydrophobic polystyrene substrate is observed for many proteins, e.g., immunoglobulins, which means that the ISFET affinity sensor

approach can be extended to applications other than heparin sensing, e.g., a possible use as ImmunoFET. Adsorption of proteins to the (highly hydrophilic) Ta_2O_5 gate oxide seems to occur less spontaneously and thus appears to require covalent modification procedures, such as chemical coupling of protein to a silylated Ta_2O_5 surface.

18. Schasfoort et al. *(5)* measured the concentration of antihuman serum albumin (HSA) antibody with an HSA-containing membrane. In a competition reaction with a charge-labeled progesterone they also measured progesterone concentration, after immobilization of antiprogesterone antibody to the membrane *(9)*. Detection limits of 10^{-7} *M* αHSA and 10^{-8} *M* progesterone were demonstrated. For surface-modified ISFETs, the theoretical detection limit for net added surface charge was calculated to be about 4×10^{10} charged groups per cm^2 when assigning significance to a differential amplitude of 0.1 mV. To give a more concrete idea this value was calculated to correspond with 0.024% of a monolayer of albumin *(10)*, which exemplifies the potential of the ion-step method for protein sensing.

References

1. Aizawa, M. (1978) Electrochemical determination of IgG with an antibody bound membrane. *J. Membr. Sci.* **4**, 221–228.
2. Yamamoto, N., Nagasawa, Y., Sawai, Y., Suda, M., and Tsubomura, T. H. (1978) Potentiometric investigations of antigen-antibody and enzyme-enzyme inhibitor reactions using chemically modified metal electrodes. *J. Immunol. Meth.* **22**, 309–317.
3. Schasfoort, R. B. M., Bergveld, P., Kooyman, R. P. H., and Greve, J. (1990) Possibilities and limitations of direct detection of protein charges by means of an immunological field-effect transistor. *Anal. Chim. Acta* **238**, 323–329.
4. Eijkel, J. C. T. (1995) Potentiometric detection and characterization of adsorbed protein using stimulus-response measurement techniques. PhD thesis, University of Twente, Enschede, The Netherlands, ISBN 90-9008615-3.
5. Schasfoort, R. B. M., Kooyman, R. P. H., Bergveld, P., and Greve, J. (1990) A new approach to ImmunoFET operation. *Biosens. Bioelectron.* **5**, 103–124.
6. Van Kerkhof, J. C., Eijkel, J. C. T., and Bergveld, P. (1994) ISFET responses on a stepwise change in electrolyte concentration at constant pH. *Sensors Actuators B* **18**, 56–59.
7. Van Kerkhof, J. C., Bergveld, P., and Schasfoort, R. B. M. (1995) The ISFET based heparin sensor with a monolayer of protamine as affinity ligand. *Biosens. Bioelectron.* **10**, 269–282.
8. Van Kerkhof, J. C., Bergveld, P., and Schasfoort, R. B. M. (1993) Development of an ISFET based heparin sensor using the ion-step measuring method. *Biosens. Bioelectron.* **8**, 463–472.
9. Schasfoort, R. B. M., Keldermans, C. E. J. M., Kooyman, R. P. H., Bergveld, P., and Greve, J. (1990) Competitive immunological detection of progesterone by means of the ion-step induced response of an ImmunoFET. *Sensors Actuators* **B1**, 368–372.
10. Van Kerkhof, J. C. (1994) The development of an ISFET-based heparin sensor. PhD thesis, University of Twente, Enschede, The Netherlands, ISBN 90-9007514-3, chapter 7.

13

Liposome-Based Immunomigration Assays

Matthew A. Roberts and Richard A. Durst

1. Introduction

During the last 10 years, there have been a number of applications of liposome reagents in various immunoassay and sensor systems. These analytical systems span a range of analyte detection for agricultural *(1)*, environmental *(2)*, and clinical interests *(3–5)*. Furthermore, similar bilayer membrane-based reagents made of red blood cell ghosts have been used for the detection of drugs of abuse in a commercially available assay *(6)*.

Previous studies have demonstrated advantages for the use of the liposome marker in competitive immunoassays *(3,7,8)*. These include the large amount of dye that can be trapped in the aqueous interior and the fact that this entrapped marker becomes immediately available for detection on liposome binding, whether through lysis of the bilayer or direct detection of encapsulated dye, as presented here. Furthermore, liposomes have been shown to have excellent long-term storage characteristics, which make them an excellent candidate for use in field-portable or point-of-care sensor systems *(4)*.

This chapter will discuss procedures, listed in flowchart format in **Fig. 1**, for the development and use of liposomes, as alternative competitive markers with the immunomigration test-strip format. A diagram of a typical test-strip is shown in **Fig. 2**. Our laboratory has previously developed the liposome-based immunomigration test strip for detection of a number of analytes, including the herbicide alachlor *(1,2,9,10)*, the natural glycoalkaloid toxins, solanine and charconine *(11)*, and the polychlorinated biphenyls (PCBs), industrial pollutants comprising a group of 209 structurally related congeners *(2)*. The methods presented in this chapter have been made general enough for application to alternative analytes that might be of more interest to the reader.

From: *Methods in Biotechnology, Vol. 7: Affinity Biosensors: Techniques and Protocols*
Edited by: K. R. Rogers and A. Mulchandani © Humana Press Inc., Totowa, NJ

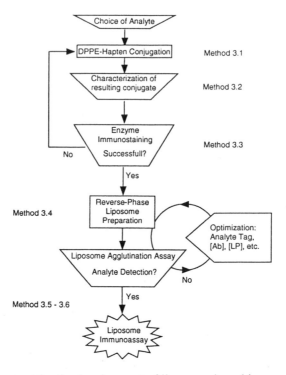

Fig. 1. Flow chart for the development of liposome-based immunosensors.

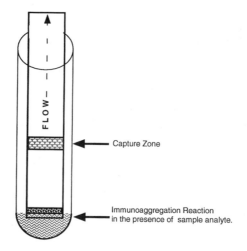

Fig. 2. Liposome-immunoaggregation sensor.

In an aqueous solution, phospholipids self-assemble to form a spherical lipid bilayer membrane that surrounds an aqueous cavity, producing a structure

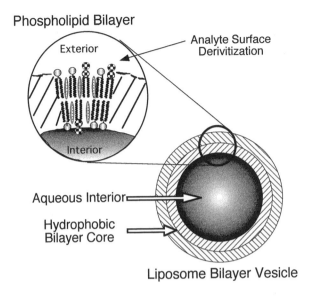

Fig. 3. Representative liposome structure.

known as a liposome. Phospholipid molecules are composed of a polar head group, and a hydrophobic tail, which consists of a double hydrocarbon chain backbone. The basic structure of liposomes and the bilayer structure can be seen in **Fig. 3**. The formation of the liposome, like the formation of micelles, is because of the fact that it is energetically favorable to form associations with the hydrophobic portion of molecules in order to reduce the interaction with a surrounding aqueous environment. The hydrophilic head groups take up more packing space than do single hydrocarbon chains (i.e., monoacyl phospholipids, detergents, lysolecithin, and many other amphipathic molecules). Thus, molecules with a single acyl chain will take on a three-dimensional cone shape, whereas those with a double chain will form a more even tubular shape. This difference drastically alters the packing arrangements they will assume when placed in aqueous solvents. Single-chain molecules most often form micelles and those having double chains form bilayer structures, such as liposomes *(12)*.

Because of the fact that phospholipid bilayer membranes spontaneously form from dispersions of phospholipids in aqueous media, the critical step in the preparation of the liposome marker is not the formation of the bilayer membranes, but of forming bilayer vesicles possessing the desired size, stability, surface characteristics, and encapsulation efficiency. The preparation of liposome types of very different characteristics may require completely different methods of preparation; however, most of the available methods have some steps in common *(12,13)*. These include the evaporative concentration of phospholipids from organic solvent, dispersion into an aqueous media, purification of organized bilayer vesicles from remaining components, and characterization of the resulting liposome population. The principal difference between the available methods is the technique used to disperse phospholipids in the aqueous medium, which can involve physical agitation, two-phase dispersion, or detergent solubilization. The reverse-phase method outlined in this chapter uses two-phase dispersion.

It is usually necessary to incorporate a detection marker with liposomes in order to be useful in a sensor system. There are numerous markers available for encapsulation in liposome bilayers, many of which can be obtained commercially. Optically detected markers include a large number of water-soluble dyes measured by their absorbance or fluorescence properties, as well as enzymes capable of converting a substrate outside of the liposome into a detectable product measured with standard, chemiluminescent, and bioluminescent detectors *(13)*. It should also be mentioned that a number of membrane probes with absorbance and fluorescence properties are commercially available for direct insertion into the membrane and subsequent detection, thereby avoiding any possible marker leakage that can be associated with use of the aqueous compartment *(14)*. Furthermore, electron spin labels, electrochemical markers, and ions may also be encapsulated for subsequent detection *(15)*. This wide range of detection options have made liposomes attractive as an alternative marker system for a wide variety of competitive immunoassay formats.

Liposomes have two distinct domains that may be altered chemically: the bilayer membrane and the interior aqueous cavity. Hydrophobic and amphipathic components can be inserted into the bilayer structure or attached to the surface. The interior aqueous cavity can be used to entrap and carry water-soluble chemicals that may be detected directly or when released after liposome lysis. The contents of this cavity may also influence the size and stability of liposomes.

For receptor-binding applications, liposomes must have a surface derivatized with one component of a ligand–receptor couple, as shown in **Fig. 3**. This can be the analyte or a structurally related analog, a nonanalyte-specific molecule, such as biotin or avidin, or an analyte receptor, e.g., antianalyte antibody, directly attached to the lipid bilayer. One benefit of using the liposome label is the high degree of control one has in the selection of the phospholipid composition. Because of the wide variety of phospholipids available, the method developer may choose from a menu of phospholipid head groups with different functional groups for conjugation to an appropriate receptor or ligand. It should be noted that when discussing phospholipid derivitization, the term analog will be used here to refer to a small molecule that is either the analyte itself or a structurally related molecule that is capable of binding to the antibody receptor with high affinity.

There are two general approaches to the surface derivatization of liposomes *(16)*. In the first case, a molecule, usually an analog, biotin, or other small molecule, is first coupled to a phospholipid carrier, which is subsequently incorporated into the bilayer during vesicle preparation. In the second case, a molecule, usually a large protein receptor, is attached to the outside of pre-formed, precharacterized liposome bilayers that have the appropriate functional group. This chapter will describe two techniques for the former method, which produce competitive markers that can be used with a number of immobilized and soluble receptor systems. These techniques involve amide or thioether linkage with dipalmitoyl phosphatidyl ethanolamine (DPPE) for analogs possessing either a carboxylic acid or chloroacetamide functional group, respectively. The resulting conjugates are added to the phospholipid mixture and incorporated into the bilayer during vesicle preparation.

The chemistry used for the coupling process is usually not specific to liposomes but has been previously developed for attachment to proteins, either as enzyme label tags or as immunogens. One common technique is to couple biotin to the terminal amino group of DPPE via the *N*-hydroxysuccinimide ester *(4,13,17)* as has been done with proteins, enzymes, and so forth. This technique is used not only in immunoassays but also in immunohistochemistry and is common enough that the DPPE-biotin conjugate is now commercially available, which greatly simplifies sensor development. The liposomes described here incorporate a DPPE-biotin conjugate during preparation at 0.1 mol% of the total bilayer components. This ligand can be used as a nonanalyte-specific binding site with either avidin or antibiotin antibody. Biotinylation of the liposome surface is useful for capturing and quantifying liposomes during immunomigration. Those liposomes not antibody bound during competition with analyte will become bound to an antibiotin zone, as shown in **Fig. 2**, where they will concentrate before quantitation. This technique is used for obtaining a liposome signal that is directly proportional to analyte concentration from a competitive immunoassay mechanism.

Liposomes have previously been shown to form immune complexes in relationship to analyte concentration *(18,19)*, as shown in the general reaction scheme of **Fig. 4**. These methods are referred to as agglutination or aggregation assays. They usually involve destabilization and lysis of contact-sensitive liposomes and are either inhibited by, or dependent on, the presence of analyte, according to the nature of the assay. Some liposome aggregation formats use complement-mediated lysis; however, this adds considerable complexity and cost to the method *(15)*. Detection for many of these lytic formats involves a fluorescence measurement of liposome released markers, and because it requires sophisticated optical equipment, these approaches have not been suitable for field analysis.

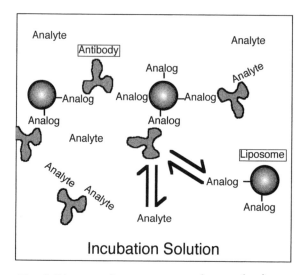

Fig. 4. Liposome immunoaggregation mechanism.

Liposome-based sensors based on aggregation reactions have been shown to be fast, simple to use, and can be evaluated visually. Both PCBs and the herbicide alachlor have been detected in the low parts per billion range *(1,2)*. Test-strips are produced from easily obtainable plastic-backed nitrocellulose, and all reagents can remain at room temperature for extended periods of time. Therefore, very little capital cost is required to produce sensors made from these components. Additionally, the immunomigration devices presented here require no changes of solution and results are generally available within 15 min.

The dye intensity may also be quantified using computer-scanning technology. The colorimetric end points can be detected visually for analyte concentrations in the low parts per billion range; however, when large numbers of sensors are being read, it is advantageous to use computer scanning to archive and then subsequently quantitate the liposome signal. Small portable scanners are now readily available that can be coupled to laptop computers and used for quantitation of liposome-based assays under nonlaboratory settings.

A flowchart for the development, production, and use of liposome-based immunomigration test-strips is shown in **Fig. 1**. This series of methods can further be grouped into three broad categories: phospholipid derivitization (**Subheadings 3.1.–3.3.**), liposome preparation (**Subheading 3.4.**), and assay optimization (**Subheadings 3.5. and 3.6.**). This chapter will take the reader step by step through the methods developed for alachlor and PCB testing with liposome-based immunomigration assays. The methods and notes presented here are written so as to suggest ways of using these procedures for alternative analytes that are of interest to the reader. The protocols listed in **Fig. 1** assume

Table 1
Reagent Sources

Reagent	Source	Location
Dipalmitoyl phosphatidyl choline (DPPC), dipalmitoyl phosphatidyl glycerol (DPPG)	Avanti Polar Lipids, Inc.	Alabaster, AL
Protein assay dye reagent, goat antirabbit IgG–alkaline phosphatase conjugate, alkaline phosphatase substrates	Bio-Rad Laboratories	Hercules, CA
Sulforhodamine B (SfB)	Eastman Chemical	Rochester, NY
N-(6-(biotinoyl)amino)hexanoyl-dipalmitoyl phosphatidyl ethanolamine (Biotin-x-DPPE)	Molecular Probes	Eugene, OR
Carnation nonfat dry milk powder (CNDM) can be obtained over the counter at a local grocery store.	Over the counter	locally available
N,N'-dicyclohexylcarbodiimide (DCC), N-hydroxysuccinimide (NHS)	Pierce	Rockford, IL
Dipalmitoyl phosphatidyl ethanolamine (DPPE), cholesterol, Tween-20, triethylamine, tris(hydroxymethyl)aminomethane (Tris), molybdenum blue spray reagent, polyvinylpyrrolidone (M_R = 40 kDa, PVP), gelatin, Sephadex G-50, n-octyl-β-D-glucopyranoside (OGP)	Sigma	St. Louis, MO
Plastic-backed nitrocellulose membranes (pore sizes >3 μm)	Schleicher & Schuell, Inc.	Keene, NH
Flexible reverse-phase silica-gel TLC plates with fluorescent indicator	Whatman	Maidstone, UK

the availability of antianalyte antibody before this development process. The production of quality antibodies for immunodiagnostic testing is too broad a subject to adequately cover in this chapter, so the reader is referred to our previous work if this is necessary *(1,2,10,14)*. All other steps and procedures are given in sufficient detail for replication by others.

2. Materials

The sources for reagents used in the following procedures are found in **Table 1**. Some common reagents and alternate sources are not listed.

2.1. Solvent Systems for Phospholipid-Analog Conjugations

1. Solvent system (SS #).
2. SS1: Chloroform:methanol; vol = ratio 1:1.
3. SS2: Chloroform with 0.7% triethylamine, preheated to 45°C.
4. SS3: 30 m*M* hydroxylamine in methanol, pH 8.2.
5. SS4: CHCl₃:MeOH:acetone:H₂O; vol ratio 65:15:15:1.
6. SS5: CHCl₃:MeOH:H₂O; vol ratio 80:22:1.2.

7. SS6: chloroform:methanol:acetone:glacial acetic acid:water; vol ratio 60:20:20:5:4.
8. Deprotection reagent: 30 mM hydroxylamine hydrochloride in methanol, adjusted to pH 8.2 with NaOH.

2.2. Identification of Analog–DPPE Conjugate by Thin-Layer Chromatography (TLC), Enzyme-Immunostaining, and Chemical Spraying

1. Silica-gel TLC plate on polyester backing, with fluorescent indicator.
2. Blocking reagent: 3.0% by weight nonfat dry milk in 20 mM Tris-HCl-buffered saline, pH 7.0, with 0.01% sodium azide.
3. Washing solution: 20 mM Tris-HCl-buffered saline, pH 7.0, with 0.05% Tween-20.
4. Antianalyte antibody solution: Rabbit antianalyte antibody at 20 µg/mL in 20 mM Tris-HCl-buffered saline, pH 7.0, containing 0.02% bovine serum albumin (BSA) and 0.05% Tween-20.
5. Goat antirabbit IgG–alkaline phosphatase conjugate solution; Goat antirabbit IgG–alkaline phosphatase (AP) conjugate diluted 1 to 3000 (v/v) in the same buffer as for the primary antibody.
6. Color developing reagent: AP conjugate substrate kit from Bio-Rad consisting of AP color reagents A and B. Reagent A contains nitroblue tetrazolium in aqueous dimethylformamide (DMF) with magnesium chloride and reagent B contains 5-bromo-4-chloro-3-indolyl phosphate in DMF.
7. Molybdenum blue spray reagent: 1.3% molybdenum oxide in 4.2 M sulfuric acid diluted with equal volume of 4.2 M sulfuric acid.

2.3. Reverse-Phase Marker-Filled Liposome Preparation

1. SS7: CHCl$_3$:isopropyl ether:MeOH (6:6:1 by volume).
2. TB: 20 mM Tris-HCl, pH 7.0.
3. TBS: Tris-HCl buffer with saline (20 mM Tris-HCl, 100 mM NaCl (only in TBS), 0.01% sodium azide, pH 7.0).
4. Marker solution: 150 mM sulforhodamine B (SfB) in 20 mM Tris-HCl, pH 7.0:
   a. Weigh 838 mg of SfB and 24.2 mg of Tris base into a graduated cylinder.
   b. Add 1 mL of TB buffer and vortex thoroughly.
   c. Bring to final volume of 10 mL with distilled H$_2$O and revortex.
   d. The resulting solution should already be close to pH 7.0; however, this should be adjusted if necessary.
5. Lipid mix: prepare the following bilayer membrane components: DPPC, cholesterol, DPPG, analog-DPPE conjugate, and biotin-x-DPPE; molar ratio of 5:5:0.5:0.1:0.01.
   a. Weigh 29.6 mg DPPC (mol wt 734.05, 40.3 µmol), 3.1 mg DPPG (mol wt 744.96, 4.2 µmol), and 15.8 mg cholesterol (mol wt 386.7, 40.9 µmol) into a 50-mL round-bottom flask with ground-glass opening.
   b. Add the analog-DPPE, dissolved in SS1, to lipid mix at 0.1 mol% (0.081 µmol) in the case of alachlor-DPPE, and 1 mol % (0.81 µmol) in the case of

4-(2-chlorophenyl)-benzoic acid (2ClPB-DPPE). Add the biotin-x-DPPE, dissolved in SS1, to lipid mix at 0.1 mol% (0.081 µmol).
5. Polycarbonate membranes: Preassembled membranes in syringe-filter format may be purchased from Poretics (Livermore, CA). For liposomes of roughly 0.3-µm diameters filters can be stacked in the following order: 3, 0.4, and 0.2 µm.
6. Size exclusion chromatography column:
   a. Equilibrate Sephadex G-50 matrix with TBS; degas just before use.
   b. Pack column (1.5-cm id × 25-cm length) and run TBS for at least 3 column volumes at roughly 3 mL/min before applying liposome preparations.
7. Lysis solution: 930 mM n-octyl-β-D-glucopyranoside in TBS. This solution may be stored at room temperature and used indefinitely.

2.4. Sensor Preparation

1. Membrane prewetting: nitrocellulose membrane is cut into 8- × 15-cm sheets, thoroughly wetted with 10% methanolic TBS, pH 7.0, and dried under vacuum for 1 h.
2. Nitrocellulose blocking solution: 2% polyvinylpyrrolidone, 0.02% gelatin, and 0.002% Tween-20 in TBS.
3. Liposome capture solutions: Antibiotin antibodies, 1 mg/mL in TBS.

2.5. Liposome-Immunoaggregation (LIA) Sensor Operation

1. Antibody solution: Antianalyte antibodies, roughly 0.2 mg/mL in TBS.
2. Liposome solution: liposome stock diluted to roughly 10^4 liposomes/µL.
3. Densitometry measurements of resulting capture zones were analyzed by Scan Analysis software (Biosoft, Ferguson, MO).

3. Methods

The following procedures are organized in sequence according to the flow chart shown in **Fig. 1**. Notes immediately follow each of the listed procedures.

3.1. Phospholipid-Analog Conjugations

3.1.1. Method for Amide Linkage (see **Notes 1–4**)

1. Analog activation: 20 µmol of carboxylated-analog are activated overnight in a small volume (approx 1 mL) of SS1 with 40 µmol N,N'-dicyclohexylcarbodiimide (DCC) and 40 µmol N-hydroxysuccinimide (NHS) while being stirred at room temperature (*see* **Notes 2 and 3**).
2. The activation solution is evaporated to dryness under a stream of N_2.
3. Analog-DPPE conjugation: 5 mg of DPPE is dissolved in SS2. This solution is used to solubilize the activated analog solution from **step 1** and then is stirred at 45°C overnight.
4. At this point confirmation of a successful conjugation reaction is desirable (*see* **Fig. 1**). This can be ascertained with a combined thin-layer chromatographic and

an enzyme-immunostaining method if an antianalog antibody is available (*see* **Subheading 3.2.**).

5. The proposed product is purified on a preparatory silica-gel plate using SS4. The purified analog-DPPE conjugate is scraped from the TLC plate into a suitable container and extracted with SS5 at 45°C.
6. SS5 is allowed to saturate the silica gel particles for 1.5 h with frequent vortexing.
7. The silica-gel–solvent is then filtered, through glass wool and a 0.45-μm polytetrafluroethylene (PTFE) filter, using a glass syringe (*see* **Note 4**).
8. The remaining silica gel is subsequently washed 3 times with 20 mL of the heated extraction solvent to ensure maximum recovery. The filtrate is taken to dryness with a rotary evaporator to ensure removal of all solvent.

3.1.2. Method for Thioether Linkage (see **Note 5**)

1. Phospholipid activation: Twenty mg of DPPE is suspended in 3 mL of SS2 and sonicated under nitrogen for 1 min. in a 45°C bath.
2. To the DPPE, 2 molar equivalents of *N*-succinimidyl-*S*-acetylthioacetate (SATA) in 1 mL SS2 is added slowly. The reaction flask is capped and stirred at room temperature for ca. 20 min; the end point of the reaction is indicated when the mixture becomes clear.
3. All solvent used in **step 2** is removed on a rotary vacuum evaporator (*see* **Note 6**).
4. Deprotection: 2 mL deprotection reagent is added, the reaction mixture is vortexed vigorously, and then stirred at 45°C for 1 h under nitrogen, maintaining the pH at 8.2 using dilute NaOH in methanol (*see* **Note 6**).
5. Conjugation: A 2.8 molar excess (to DPPE) of alachlor in 1 mL of SS3 is added to the reaction flask. The reaction mixture is stirred at 45°C for 2 h, with the pH being maintained at 8.2. The reaction is then continued at 45°C overnight (*see* **Note 7**).
6. The proposed product is purified on a preparatory silica-gel plate, as described for **Subheading 3.1.1.**; however, SS6 is used (*see* **Note 8**).

3.2. Identification of Analog-DPPE Conjugates by Thin-Layer Chromatography Followed by Enzyme-Immunostaining and Chemical Spraying Method (see Note 9)

3.2.1. TLC-Enzyme-Immunostaining Method (see **Note 10**)

1. Spot the proposed analog-DPPE reaction mixture in duplicate on a reversed-phase TLC plate and develop in SS4 or SS6, i.e., SS4 for an amide-linked conjugate and SS6 for a thioether-linked conjugate.
2. The resulting TLC plate is cut into two equal parts with a standard glass-cutting tool. One half of the chromatographed plate will be observed using enzyme-immunostaining in this procedure and the other half will be visualized using spray detection in **Subheading 3.2.2.**
3. Dry one-half plate in a vacuum oven at room temperature for 2 h.

4. Place the dry TLC plate in blocking reagent on a rotary shaker (gentle motion) for 1 h at room temperature.
5. Wash the blocked TLC plate three times (5–10 min each time) with the washing solution.
6. Incubate the plate with the antianalog antibody solution overnight at room temperature with constant gentle shaking.
7. Wash the plate as in **step 4** above.
8. Incubate the plate with the goat antirabbit IgG–alkaline phosphatase solution for 2 h at room temperature on the shaker.
9. Wash the plate as in **step 4**.
10. Immerse the plate in the color development solution. Incubate at room temperature, with gentle agitation, until color development is complete.
11. Stop the color development by washing the plate in distilled water for 10 min and let the plate air dry.

3.2.2. TLC-Spray Detection Method (see **Note 12**)

1. After air-drying, one half of the TLC plate, prepared in **Subheading 3.2.1.**, is examined under a UV lamp (254 nm). Mark all visible spots.
2. Place the TLC plate in a sealed chamber saturated with iodine vapor. Mark all reddish brown spots that appear.
3. Spray the plate with the molybdenum blue spray reagent. A positive blue stain indicates the presence of phospholipids. Mark all blue spots (*see* **Note 11**).

3.3. Reverse-Phase Evaporation Technique for Liposome Preparation (see *Notes 13 and 14*)

1. Dissolve lipid mix in 8 mL SS7.
2. Sonicate at 45°C under N_2 for 1 min to effect total dissolution of lipid components.
3. Add 1.4 mL marker solution.
4. Continue to sonicate at 45°C under N_2, while gently swirling flask, for 5 min.
5. Remove SS7 with a rotary vacuum evaporator attached to a water aspirator, keeping the vacuum between 19 and 25 psi. Water bath temperature should be 4°C. Liposome preparation will begin to foam after 1–2 min, however, vacuum pressure is continued just until foaming stops. Remove flask from evaporator (*see* **Note 15**).
6. Add an additional 2.6 mL marker solution to the flask, swirl gently, and resonicate at 45°C under N_2 for 3 min (*see* **Note 16**).
7. Pass the resulting solution three times through a series of polycarbonate-membrane syringe filters (decreasing pore sizes down to the desired liposome size range) (*see* **Note 17**).
8. Separate unencapsulated marker from intact marker-loaded liposomes on a size exclusion chromatography column. Recover the liposome fraction directly after void volume elution (*see* **Note 18**).
9. Dialyze the recovered liposomes in 1 L of TBS at 4°C in the dark. Change buffer at least twice before storing liposomes in Eppendorf tubes at 4°C in the dark.

10. Determination of liposome integrity: dilute a small aliquot of the stock liposome preparation 5000 times in TBS. Measure the fluorescence of this solution before and after addition of lysis solution (33 μL lysis solution/1 mL 1:5000 liposomes). SfB is excited at 543 nm and the emission is measured at 596 nm (*see* **Notes 19–21**).

3.4. Sensor Preparation

1. A prewetted 8- × 15-cm nitrocellulose membrane section is mounted on a microprocessor-controlled TLC sample applicator (*see* **Notes 22** and **23**).
2. Desired capture solutions are applied 2 cm from the bottom at 1.25 μL/s for 85 s with 190 kPa (27.5 psi) N_2, producing discrete zones as shown in **Fig. 2**. Sheets are then vacuum dried for 1.5 h (*see* **Note 24**).
3. The coated nitrocellulose sheet is immersed in blocking solution for 1 h on a rotating mixer and dried under vacuum for 3–4 h.
4. Prepared sheets are cut into 5- × 80-mm sensor strips, vacuum packed (standard food sealer found in grocery stores may be used), and stored in the presence of silica-gel desiccant at room temperature until ready for use (*see* **Note 25**).

3.5. Liposome-Immunoaggregation (LIA) Sensor Operation (see Notes 26–32)

1. The format for the LIA assay, shown in **Fig. 2**, consists of a solution containing the sample to be analyzed, analyte-tagged liposomes, antianalyte antibodies, and a nitrocellulose test-strip with an immobilized capture zone.
2. The assay is initiated by dispensing 25 μL of the sample (in water or up to 30% methanol extraction solvent) and 25 μL of antibody solution into a 10- × 75-mm glass test tube.
3. The solution from **step 2** is shaken gently to ensure adequate mixing, and 25 μL of the liposome solution is then added, again with gentle shaking.
4. After mixing, the solution is allowed to incubate at room temperature for approx 5 min before continuing with the assay. After these initial preparations the test tube is shaken mildly to mix the contents and the test-strip is inserted.
5. The test-strip is left in the tube until the solution front reaches the end of the strip (approx 8 min); the strip is then removed and air dried. The color intensity of all zones is estimated visually or by scanning densitometry.

4. Notes

4.1. Amide Linkage

1. This method was modified from the procedure of Schmidt et al. *(20)* to couple a carboxylated PCB analog, 2ClPB, to the phospholipid carrier, DPPE, for subsequent insertion into liposome bilayers *(2)*. The complete reaction scheme is shown in **Figs. 5** and **6**. The chemistry used here has been widely used in the literature for various applications and should be amenable to most analytes with carboxylic acid functional groups.

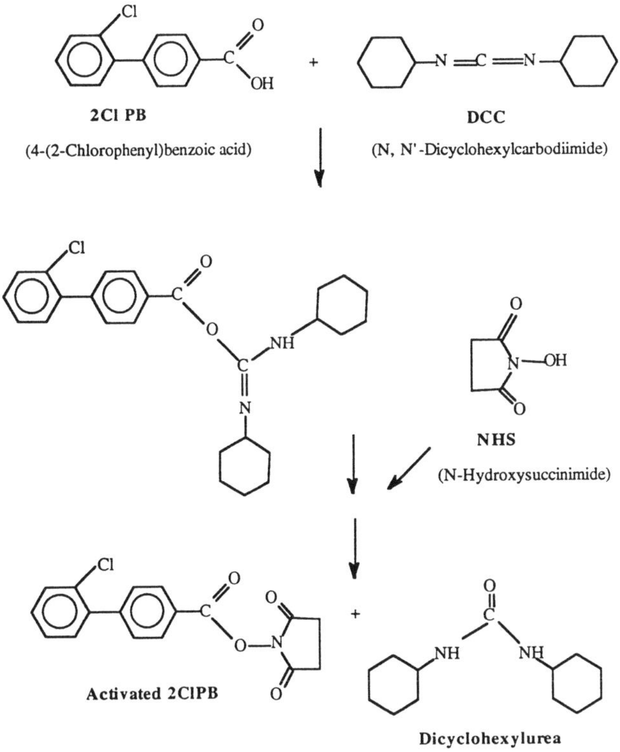

Fig. 5. 2ClPB activation reaction.

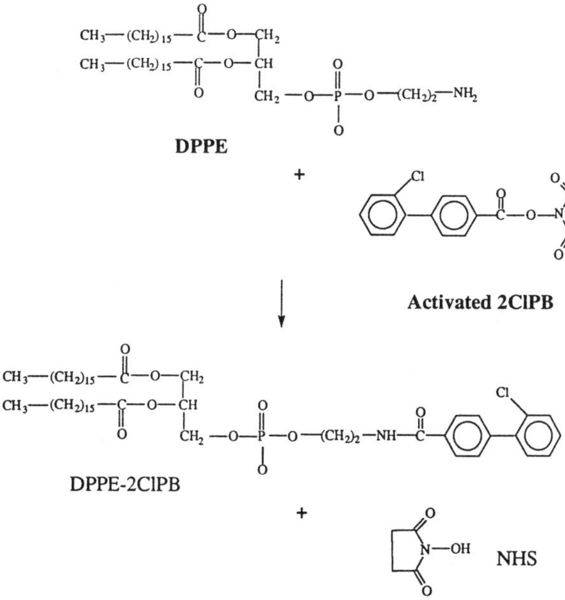

Fig. 6. DPPE conjugation with activated 2ClPB.

2. If the analyte is significantly hydrophilic then SS1 will need to be modified accordingly.

3. A general scheme for the formation of the *N*-hydroxysuccinimide ester from the carboxylated-analog, in this case 2ClPB, is shown in **Fig. 5**. As can be seen in this figure, a byproduct, dicyclohexylurea, is formed, which precipitates out of the reaction mixture. It should be noted that the presence of a precipitate is not necessarily limited to this byproduct and may contain analog or analog-conjugate as well. The method presented here produces usable quantities of activated 2ClPB that remain soluble in SS1. If this solvent system is changed to accommodate an alternate analyte of interest then this yield may change because of precipitation at this reaction step. The enzyme immunostaining method presented for characterization of the final analog-DPPE conjugate will determine if the conjugate has been produced. If the proposed conjugate fails this test then **step 1** of **Subheading 3.1.1.** should be repeated with the concomitant analysis of the reaction mixture at various time points by TLC or HPLC to determine if, in fact, a new product is being formed.

4. If the resulting conjugate is not to be immediately incorporated into liposome bilayers storage at –20°C is recommended.

4.2. Thioether Linkage

5. This method was developed for conjugation of the herbicide, alachlor, to the phospholipid carrier DPPE (alachlor-DPPE) by modifying the thiolation procedure of Feng et al. *(21)* to include the use of the SATA reagent as developed by Duncan et al. *(22)*. The analyte or analog will be referred to here specifically as alachlor, however, the procedure should be amenable to other analytes with a chloroacetamide functional group.

6. After the activation reaction has proceeded to completeness, the product of interest, ATA-DPPE, contains an *S*-acetylthioacetylate group capable of forming a thioether linkage, as shown in **Fig. 7**. Before this linkage can be formed, it is necessary to first carry out a deprotection step to expose a sulfhydryl group, as shown in **Fig. 8**. Once deprotected in **steps 3** and **4**, **Subheading 3.1.2.**, the sulfhydryl group becomes capable of thioether formation with the analyte, alachlor.

7. After completion of this step, alachlor-DPPE is formed, as shown in **Fig. 9**.

8. Referring to method **step 6**: If the resulting conjugate is not to be immediately incorporated into liposome bilayers, then storage at –20°C is recommended.

4.3. Analog-DPPE Conjugate

9. To summarize, this procedure involves chromatographing the reaction mixture in an appropriate thin-layer chromatographic plate and solvent system in duplicate. One of the TLC plate halves is dried, blocked, washed, and then incubated overnight in a solution that contains antibody to alachlor. This plate is washed and placed in a solution containing a goat antirabbit–IgG-alkaline phosphatase conjugate for 2 h. The plate is again washed and developed with the substrate for alkaline phosphatase. When color development is complete, the plate is washed

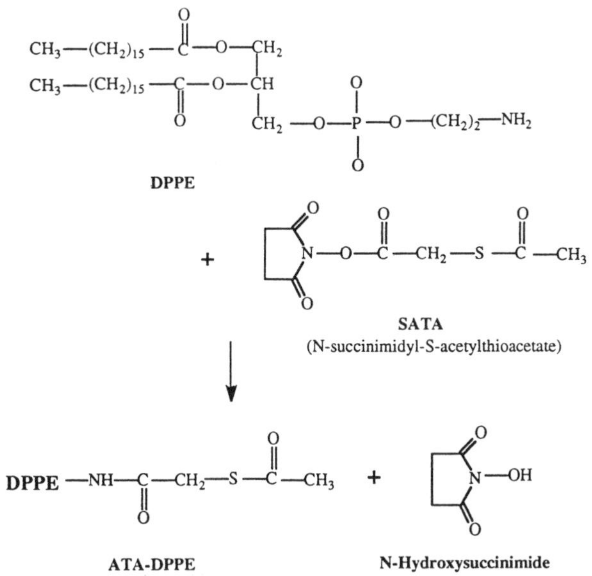

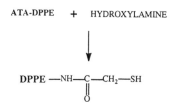

CH_3—$(CH_2)_{15}$—$\overset{\overset{O}{\|}}{C}$—$O$—$CH_2$

CH_3—$(CH_2)_{15}$—$\overset{\underset{O}{\|}}{C}$—$O$—$CH$

$\qquad \qquad \qquad CH_2$—$O$—$\overset{\overset{O}{\|}}{\underset{O}{P}}$—$O$—$(CH_2)_2$—$NH_2$

DPPE

+

$\overset{O}{}$ N—O—$\overset{\overset{O}{\|}}{C}$—$CH_2$—$S$—$\overset{\overset{O}{\|}}{C}$—$CH_3$

SATA
(N-succinimidyl-S-acetylthioacetate)

↓

DPPE —NH—$\overset{\underset{O}{\|}}{C}$—$CH_2$—$S$—$\overset{\overset{O}{\|}}{C}$—$CH_3$ + N—OH

ATA-DPPE   **N-Hydroxysuccinimide**

Fig. 7. DPPE-SATA activation chemistry for alachlor conjugation.

ATA-DPPE + HYDROXYLAMINE

↓

DPPE —NH—$\overset{\underset{O}{\|}}{C}$—$CH_2$—SH

Fig. 8. Deprotection of ATA-DPPE by hydroxylamine.

in distilled water and dried. A purple spot indicates the presence of alachlor. The other TLC plate half is examined under the UV lamp (254 nm), exposed to iodine vapor, and then sprayed with the molybdenum blue reagent, which is specific for phospholipids. Various reaction substrates and byproducts can be observed with the UV light and the iodine stain, which may be used for monitoring the progress of reactions. The alachlor-DPPE spot should appear purple with the alkaline phosphatase substrate stain and blue with the molybdenum blue spray reagent.

10. This method inherently assumes the availability of antianalyte antibody; however, if this is not the case then other options are available. If a radiolabeled analog is available, then this may be conjugated as previously described, run under identical TLC conditions, and then imaged using autoradiography. Secondly, if the TLC conditions do not provide adequate resolving power then a technique with a higher number of theoretical plates may be used, such as high-performance liquid chromatography (HPLC).

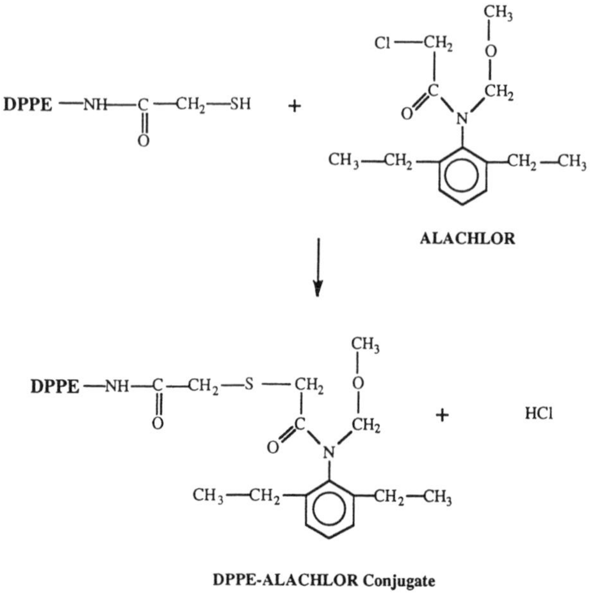

Fig. 9. Alachlor-DPPE conjugation by thioether linkage.

11. In **Subheading 3.2.2., step 3** it was indicated to "mark" spots on the TLC. This can be accomplished simply by tracing the spots with a pencil. Additionally, we have found for archival purposes it is advantageous to digitize the developed plate image with a standard desktop computer scanner.

12. It should also be noted that if the conjugation is determined to be successful, quantitation of the conjugate will at some point be necessary. A molecule of DPPE has only one atom of phosphorous; therefore, any method for determination of phosphorous content can be used to determine the amount of phospholipid-conjugate. We have previously used Bartlett's method *(23)* for conjugate as well as liposome quantification. Since several techniques for phosphorous determination are available in many laboratories, the reader is encouraged to use whatever method is most readily available.

4.4. Liposome Preparation

13. The reverse-phase evaporation method presented here is a good general preparation technique. Resulting liposomes may contain large marker concentrations (up to 200 mM) and have been shown to possess refrigerated shelf-lives on the order of years *(1,2,4,10,14)*. They are also stable at room temperature as long as care is taken to cover them from excessive light exposure. Furthermore, the degree and type of analyte derivatized to the surface is easily optimized for individual assay conditions.

14. The liposome preparation step is obviously at the core of any liposome-based immunosensor development work, and, therefore, critical thought should initially

be given to outlining the desired physical properties that will be required of liposome reagents in the final analytical procedure before preparation is initiated. For example, if one wishes to develop an assay step that involves lysis at a specified temperature then the phase transition temperature (T_m) of individual lipid components must be considered. A number of phospholipids having T_ms from −10 to 70°C are commercially available and, therefore, could be chosen as alternatives to our lipid mix. The point of this note is that the preparation method listed in this chapter is one of many. A text by R. R. C. New, *Liposomes: A Practical Approach (12)*, is a good starting place if one wishes to explore liposome preparation with other desired characteristics.

15. In **Subheading 3.3.**, **step 5**, care must be taken not to apply too much vacuum pressure to speed up the evaporation process. The change from a foaming to a bursting process can be very rapid and loss of the liposome preparation is likely above 25 psi. A trap can be installed between the round-bottom flask and the evaporator; however a liposome burst will reduce yield considerably even when taking this precaution.

16. In **Subheading 3.3.**, **step 6**, if an oily film has been deposited on the top of the round-bottom flask, then vortexing may be necessary as well to bring these components back into the aqueous phase where liposomes are formed.

17. In **Subheading 3.3.**, **step 7**, the size and heterogeneity of the liposomes can essentially be chosen by selecting the pore size of the polycarbonate membranes. Liposomes will tend to form spheres with diameters slightly large than the membrane pore size because they are able to deform slightly under pressure and squeeze through a restricted space. After preparation, liposome size will be distributed about some mean with a deviation characteristic for the method. For this method liposomes will be produced with diameters of roughly 0.3 ± 0.09 μm for a single pass through the series of filters described. This deviation can be reduced by increasing the number of passes through the filter set. In fact, there are commercially available instruments that will automatically pass liposome preparations through polycarbonate membranes for hundreds or thousands of passes to obtain the tightest possible size distribution. However, for use with liposome-based immunomigration test-strips, a relatively large size distribution will not drastically affect performance.

18. In **Subheading 3.3.**, **step 8**, only about 3–10% of the marker solution is actually encapsulated within liposome bilayers. It should be noted that the two fractions, intact liposomes and unencapsulated marker, are easily distinguishable on the size exclusion chromatography column. The SfB marker preparation is made at a very high concentration at which its native fluorescent properties are quenched and the solution exhibits a very dark, almost black, coloration. SfB not encapsulated in liposomes will be slightly diluted on the size exclusion column and become brightly colored. Therefore, the intact liposomes that encapsulate SfB marker are easily identified as the very dark, almost black, zone.

19. Because of the fact that SfB is highly fluorescent and this fluorescence undergoes self-quenching when encapsulated, the integrity of liposome preparations

Table 2
Typical Reverse-Phase Liposome Characteristics[a]

Mean diameter (± 1 SD)	0.3 ± 0.09 μm
Volume	1.4×10^{-11}/μL
Liposome concenttration	2.1×10^9 liposomes/μL
SfB concentration, liposomal	150 mM
SfB content	1.2×10^6 molecules/liposome
Analog surface density (0.1–1 mol%)	10^2–10^3 molecules/liposome
Biotin surface density (0.1 mol%)	10^2 molecules/liposome
Stability	95% intact after 1 yr

[a]Measurements performed on liposomes stored at 4°C in TBS.

can be determined by measuring fluorescence intensity before and after lysis. Total and almost instantaneous lysis of the liposomes is effected by addition of 930 mM n-octyl-β-D-glucopyranoside. The ratio of the fluorescent signal after after lysis to before lysis (encapsulated versus unencapsulated marker) should always be greater than 20 for well-prepared liposomes.

20. The liposome integrity should be monitored versus time, especially when the reagent is used with new types of sample matrices. Liposomes with encapsulated SfB tend to be robust shelf-stable reagents but can lyse under conditions of intense light and high concentrations of organic solvents.

21. The listed method will yield liposomes with roughly the characteristics listed in **Table 2**. For more precise measurements of liposome size and concentration, laser-light scattering particle size analysis or electron microscopy may be performed *(1)*.

4.5. Notes on Sensor Preparation

22. The membrane selected for sensor preparation has a serious impact on immunoassay performance. Variations in membrane adsorptivity and wetting is probably the key negative issue in liposome-based immunosensors that cannot be controlled within the laboratory and yet must be dealt with to obtain reproducible results. This variation can be especially large among manufacturers and even from lot to lot within manufacturers. In our laboratory, nitrocellulose obtained from Schleicher and Schuell, Millipore, and the Sartorious Co. were found to give adequate assay performance.

23. Fortunately, most serious variations affecting sensor performance can be detected during the membrane prewetting stage (*see* **Subheading 2.4.**). When properly wetted the membrane will turn from white to an off-gray. At this stage it is typical to see several spots that remain white even after being immersed in the 10% methanolic TBS. The location of these spots should be noted. After drying, an 8-cm long strip should be cut out of the 8- × 15-cm piece so as to remove any potential sensor strips that would contain these spots. If this step is taken then immunoassay variability can be limited to roughly 5% residual standard deviation.

24. For **Subheading 3.4.**, **step 2**, our laboratory has relied on a Linomat IV (CAMAG Scientific Inc., Wrightville Beach, NC) for application of the capture zone. The key criteria required for such a device are reproducible spraying of protein-based solutions in narrowly defined zones (preferably <5 mm) and constant table motion at roughly 10 mm/s. The device thereby sprays receptor solutions onto nitrocellulose sheets moving underneath the spray nozzle at a constant rate.

25. It should be noted that this procedure produces a generic solid support. That is, any biotinylated liposome preparation may be quantitated in the antibiotin capture zone. Any liposomes derivatized with a suitable analyte analog may undergo competition with analyte for antibody-binding sites in a homogeneous incubation solution. The resulting molecular aggregates are then separated by the nitrocellulose polymer structure, and the relatively unbound liposomes will be trapped or quantitated at this single simply prepared binding zone.

4.6. Notes on Sensor Operation

26. The critical step in this technique is the homogeneous incubation reaction between sample analyte, antibody, and derivatized liposomes as shown in **Fig. 4.** In the absence of analyte in the incubation solution, the liposome reagent will undergo multiple binding reactions with the antibody receptor. The greater the concentration of sample analyte in the incubation solution the more this process will be inhibited.

27. The resulting antibody–liposome aggregates are unable to travel on the blocked nitrocellulose substrate because of increased affinity for the nitrocellulose polymer or the increased physical size of the aggregates. The binding strength or affinity between the antibody–liposome pair will control which of these two mechanisms predominate. Nevertheless, these aggregates are inhibited from migration up the test-strip and will form a dark-colored zone at the meniscus of the incubation solution that is inversely proportional to analyte concentration, as shown in **Fig. 2.** Liposomes not sufficiently antibody-bound will migrate freely until encountering the antibiotin capture zone where they will bind, forming a second colored zone that is proportional to analyte concentration (*see* **Fig. 2**).

28. The immunoaggregation reaction just described will be sensitive to the concentration of antibody, the concentration of liposomes, and the amount of derivatization of the liposome surface. Optimization of the liposome-based sensor strips for new analytes will, variables. Fortunately, after determination of a successful analog-DPPE conjugation these variables are easily manipulated by the methods developer.

29. It should be noted that after the initial incubation reaction, immunomigration on the test strips is essentially self-timing. The TBS buffer takes approx 8 min to reach the top of the strip on which color development is finished. If left unattended for long periods of time (>45 min) evaporation will lead to more buffer flow and, consequently, a deepening of the colored zones. The point here is that critical timing of immunomigration is not necessary and, therefore, parallel analysis of up to 50 test-strips may be accomplished simultaneously.

30. We have empirically found the capture zone to produce a more homogeneously colored zone than the meniscus zone. If used for quantitation, this zone results in lower standard deviations, whether using visual estimation or scanning densitometry.
31. Visual estimation of the color intensity at either zone can be used (Comparisons are made with a reference color card), but for more accurate quantitation it was found to be preferable to use a standard desktop scanner and software capable of converting images of the resulting capture zones into gray-scale density readings.
32. Parallel quantitation by scanning densitometry of hundreds of test-strips maybe performed simultaneously and is only limited by the scanning area.

4. References

1. Siebert, S. T. A., Reeves, S. G., Roberts, M. A., and Durst, R. A. (1995) Improved liposome immunomigration strip assay for alachlor determination. *Anal. Chim. Acta* **311,** 309–318.
2. Roberts, M. A. and Durst, R. A. (1995) Investigation of liposome-based immunomigration sensors for the detection of polychlorinated biphenyls. *Anal. Chem.* **67,** 482–491.
3. Locascio-Brown, L., Plant, A. L., Horvath, V., and Durst, R. A. (1990) Liposome flow injection immunoassay: implications for sensitivity, dynamic range, and antibody regeneration. *Anal. Chem.* **62,** 2587–2593.
4. Plant, A. L., Brizgys, M. V., Locasio, B. L., and Durst, R. A. (1989) Generic liposome reagent for immunoassays. *Anal. Biochem.* **176,** 420–426.
5. Plant, A. L., Gray, M., Locascio-Brown, L., and Yap, W. T. (1993) Hydrodynamics of liposomes and their multivalent interactions with surface receptors, in *Liposome Technology* (Gregoriadis, G., ed.), CRC, Boca Raton, FL, pp. 439–454.
6. Parsons, R. G., Kowal, R., LeBlond, D., Yue, V. T., Neargarder, L., Bond, L., Garcia, D., Slater, D., and Rogers, P. (1993) Multianalyte assay system developed for drugs of abuse. *Clin. Chem.* **39,** 1899–1903.
7. Yap, W. T., Locascio-Brown, L., Plant, A. L., Choquette, S. J., Horvath, V., and Durst, R. A. (1991) Liposome flow injection immunoassay: model calculations of competitive immunoreactions involving univalent and multivalent ligands. *Anal. Chem.* **63,** 2007–2011.
8. Durst, R. A., Locascio-Brown, L., and Plant, A. L. (1990) Flow injection analysis based on enzymes or antibodies, in *GBF Monograph Series: Flow Injection Analysis* (Schmid, R. D., ed.), VCH, Weinheim, pp. 181–190.
9. Reeves, S. G., Rule, G. S., Roberts, M. A., Edwards, A. J., and Durst, R. A. (1994) Flow-injection liposome immunoanalysis (FILIA) for alachlor. *Talanta* **41,** 1747–1753.
10. Siebert, T. A., Reeves, S. G., and Durst, R. A. (1993) Liposome immunomigration field assay device for Alachlor determination. *Anal. Chim. Acta* **282,** 297–305.
11. Glorio, P. (1997) Development of a liposome-immunomigration strip assay for the determination of potato glycoalkaloids. PhD dissertation, Cornell University, Ithaca, NY, 143 pp.

12. New, R. R. C., ed. (1990) *Liposomes: A Practical Approach.* Oxford University Press, New York, pp. 33–103.
13. Rongen, H. (1995) Thesis: "Immunoassays: Development and application of chemiluminescent and liposome labels" University of Utrecht, The Netherlands, 176 pp.
14. Roberts, M. A., MacCrehan, W. A., Locascio-Brown, L., and Durst, R. A. (1996) Use of liposomes as analytical reagents in capillary electrophoresis. *Anal. Chem.* **68**, 3434–3440.
15. Monroe, D. (1990) Novel liposome immunoassays for detecting antigens, antibodies, and haptens, *J. Liposome Res.* **1**, 339–377.
16. Rongen, H. A. H., Van Nierop, T., Van Der Horst, H. M., Rombouts, R. F. M., Van Der Meide, P. H., Bult, A., and Van Bennekom, W. P. (1995) Biotinylated and streptavidinylated liposomes as labels in cytokine immunoassays. *Anal. Chim. Acta* **306**, 333–341.
17. Bayer, E. A., Rivnay, B., and Skutelsky, E. (1979) On the mode of liposome-cell interactions: biotin-conjugated lipids as ultrastructural probes. *Biochim. Biophys. Acta* **550**, 464–473.
18. Pinnaduwage, P. and Huang, L. (1992) Stable target-sensitive immunoliposomes. *Biochemistry* **11**, 2850–2855.
19. Babbitt, B., Burtis, L., Dentinger, P., Constantinides, P., Hillis, L., Mcgirl, B., and Huang, L. (1993) Contact-dependent immune complex-mediated lysis of hapten-sensitized liposomes. *Bioconjugate Chem.* **4**, 199–205.
20. Schmidt, D. J., Clarkson, C. E., Swanson, T. A., Egger, M. L., Carlson, R. E., Van Emon, J. M., and Karu, A. E. (1990) Monoclonal antibodies for immunoassay of avermectins. *J. Agricult. Food Chem.* **38**, 1763–1770.
21. Feng, P. C. C., Wratten, S. J., Horton, S. R., Sharp, C. R., and Logusch, E. W. (1990) Development of an enzyme-linked immunosorbent assay for alachlor and its application to the analysis of environmental water samples. *J. Agricult. Food Chem.* **36**, 159–163.
22. Duncan, R. J., Weston, P. D., and Wrigglesworth, R. (1983) A new reagent which may be used to introduce sulfhydryl groups into proteins, and its use in the preparation of conjugates for immunoassay. *Anal. Biochem.* **132**, 68–73.
23. Bartlett, G. R. (1958) Phosphorous assay in column chromatography. *J. Biol. Chem.* **234**, 466–468.

14

Isolated Receptor Biosensors Based on Bilayer Lipid Membranes

Masao Sugawara, Ayumi Hirano, and Yoshio Umezawa

1. Introduction

Biosensors based on isolated receptors, together with planar bilayer lipid membranes (BLMs), provide a highly sensitive and selective sensing method for bioactive substances. Although isolation of receptors from mammals is necessary for designing the sensors, the sensors are easily fabricated, if adequate procedures for isolation of receptor proteins and formation of receptor-incorporated BLMs are made. This chapter describes two examples of biosensors based on isolated receptors: glutamate receptor ion channels (GluRs) that induce membrane permeability changes on binding L-glutamate, and a Na^+/D-glucose cotransporter that displays active transport of D-glucose.

The principles of these sensors are demonstrated in **Fig. 1**. The isolated GluRs and Na^+/D-glucose cotransporter are embedded in planar BLMs formed by the monolayer-folding method. The GluRs, upon binding L-glutamate, open the channels through which a large amount of cations ($\sim 2.1 \times 10^5$ ions/channel) are allowed to permeate, thereby inducing changes the membrane permeability (**Fig. 1A**). The flux of ions thus permeated is measured as an ion current, which is an amplified measure of L-glutamate concentration. The Na^+/D-glucose cotransporter displays active transport of D-glucose (**Fig. 1B**). The driving force for D-glucose to be pumped is provided as an electrochemical Na^+ gradient. The Na^+/D-glucose cotransporter is specifically activated by Na^+ ions, and the energy conversion proceeds in a coupled transport: a Na^+ flux, following its electrochemical gradient, is coupled to a D-glucose flux with a coupling stoichiometry of Na^+ to D-glucose either at 1:1 or 2:1. The cotransported Na^+

From: *Methods in Biotechnology, Vol. 7: Affinity Biosensors: Techniques and Protocols*
Edited by: K. R. Rogers and A. Mulchandani © Humana Press Inc., Totowa, NJ

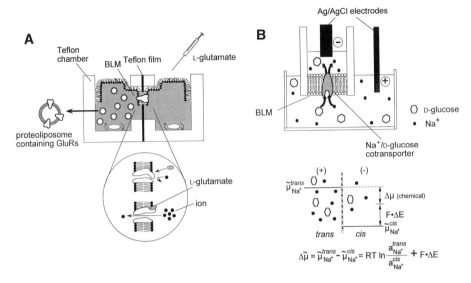

Fig. 1. Principles of biosensors based on isolated receptors. **(A)** Glutamate receptor ion channels (GluRs) and **(B)** Na^+/D-glucose cotransporter.

ions across the BLM are measured as a Na^+ ion current. The amount of this current is an analytical measure of D-glucose concentration. These ion-channel and active-transport sensors enable a highly sensitive and selective detection of L-glutamate *(1)* and D-glucose *(2)* at nM levels, respectively. Also, the chemical selectivity of GluRs toward agonists based on a physiologically relevant measure, i.e., membrane permeability changes, can be obtained with the ion-channel sensor *(3)*.

For constructing the BLM biosensors, different types of planar BLMs can be used, i.e., bilayer lipid membranes formed by the monolayer-folding method (Takagi-Montal method) *(1–3)* and micro BLMs by the tip-dip methods *(4)*. The BLMs thus formed have different sizes, ranging from ~1 μm (tip-dip method) to 20–400 μm (monolayer-folding method) in diameter. In this chapter, BLMs formed by the monolayer-folding method are exclusively used.

The procedures for isolation of two receptors, glutamate receptor ion-channel proteins (GluRs) and Na^+/D-glucose cotransporter, are described in **Subheading 3.1.** The GluRs are isolated and partially purified from whole brain of rats. Synaptic plasma membranes containing GluRs are used for sensor construction *(3)*. The GluRs may purposely be further purified for constructing the sensor *(1,4)*. The Na^+/D-glucose cotransporter is isolated and purified from small intestine of guinea pig. The preparation gives 3 polypeptide bands (67, 59, and 48 kDa), including Na^+/D-glucose cotransporter (67 kDa), on sodium dodecyl sulfate-polyacrylamide gel electrophoresis *(3)*.

The isolated receptors can be incorporated in planar BLMs by two different approaches, (1) fusion of proteoliposomes containing receptors to the BLM formed by the folding method, and (2) folding of lipid monolayers containing receptors at the air-solution interface to form receptor-incorporated BLMs. The first approach is used for incorporation of GluRs and the second approach for incorporation of Na^+/D-glucose cotransporter, as described in **Subheading 3.4.**

Subheadings 3.3. and **3.5.** describe how to set up the measuring system and how to measure the responses of the biosensors, respectively. Because the observed current is at picoamp levels, it is necessary to be set up the measuring system carefully for obtaining discrimination of analytical signals from electrical noises. The Notes for **Subheadings 3.3.–3.5.** emphasize the empirical factors to be considered for obtaining reliable analytical signals.

2. Materials

2.1. Isolation of Receptors

2.1.1. Synaptic Plasma Membranes Containing Glutamate Receptor Ion Channels (GluRs)

1. Six anesthetized adult male Sprague-Dawley rats (7 wk).
2. Buffer I: 10 mM HEPES-NaOH, pH 7.4, containing 0.32 M sucrose, 0.5 mM $MgSO_4$, 5% glycerol, 0.1 mM benzamide, 0.2 mM benzamidine-HCl, 10 mM 6-amino-n-caproic acid, 0.1 mM EGTA.
3. Buffer II: 10 mM HEPES-NaOH, pH 7.5, containing 0.32 M sucrose, 0.01 mM benzamide, 0.02 mM benzamidine-HCl, 1 mM 6-amino-n-caproic acid, 0.01 mM EGTA.
4. Ficoll A: 8% ficoll (type 400) in buffer II (17 µL of **item 6** and 34 µL of **item 7** should be added to 170 g of ficoll A just before use).
5. Ficoll B: 14 % ficoll (type 400) in buffer II (17 µL of **item 6** and 34 µL of **item 7** should be added to 170 g of ficoll B just before use).
6. Pepstatin solution: 1 mM pepstatin in ethanol (this solution can be prepared by dissolving pepstatin in hot ethanol).
7. PMSF solution: 6 mM phenylmethanesulfonyl fluoride (PMSF) in ethanol.
8. HEPES buffer: 10 mM HEPES-NaOH, pH 8.0, containing 0.25 M sucrose, 20 mM potassium acetate, 0.1 mM EDTA, 10% glycerol (167 µL of **item 9** should be added to 100 mL of HEPES buffer just before use).
9. DTT solution: 10 mg dithiothreitol (DTT) in 1 mL of water.
10. Isolation buffer I: 250 mL of buffer I plus 200 µL of pepstatin (**item 6**) plus 500 µL of PMSF (**item 7**) (these should be added just before use).
11. Isolation buffer II: 1 L of buffer II plus 56 µL of pepstatin (**item 6**) plus 140 µL of PMSF (**item 7**). (These should be added just before use.) **Items 2, 3, 4, 5, 8, 10,** and **11** should be stored at 4°C.

2.1.2. Na⁺/D-Glucose Cotransporter

1. Ten anesthetized female guinea pigs (~250 g/pig).
2. 300 MHT buffer: 10 mM HEPES/Tris, pH 7.5, containing 300 mM D-mannitol.
3. 400 MHT buffer: 10 mM HEPES/Tris, pH 7.5, containing 400 mM D-mannitol.
4. CaCl$_2$.
5. SDS solution: 2 mg/mL sodium dodecyl sulfate (SDS)
6. NaCl buffer: 50 mM K$_2$HPO$_4$, pH 7.4, containing 100 mM NaCl, 10 mM D-glucose, 10 mM L-proline.
7. Sucrose solution I: 10% sucrose in 50 mM K$_2$HPO$_4$, pH 7.4.
8. Sucrose solution II: 35% sucrose in 50 mM K$_2$HPO$_4$, pH 7.4.
9. KCl buffer : 25 mM HEPES/Tris, pH 7.4, containing 150 mM KCl.
10. 25 HT buffer : 25 mM HEPES/Tris, pH 7.4.
11. 5 HT buffer: 5 mM HEPES/Tris, pH 7.5, containing 100 μg/mL PMSF.
12. Pretreatment buffer: 20 mM imidazole/HCl, pH 7.4, containing 100 μg/mL PMSF.
13. A starting buffer: 100 μg/mL PMSF, 1% 3-[(3-cholamidopropyl)dimethyl ammonio]-1-propanesulfonate (CHAPS), 20 mM imidazole/HCl, pH 7.4.
14. A chromatofocusing column: polybuffer exchanger PBE94, 40- × 1.5–cm.
15. An elution buffer: polybuffer 74.
16. Poly(vinyl pyrrolidone) (PVP). **Items 2–15**, except for **item 4** should be stored at 4°C.

2.2. Preparation of Proteoliposomes

2.2.1. Glutamate Receptor Ion Channels

1. Synaptic plasma membrane suspension containing GluRs (it should be stored at 4°C).

2.2.2. Na⁺/D-Glucose Cotransporter

1. Purified Na⁺/D-glucose cotransporter (it should be stored under nitrogen at −80°C).
2. Tris buffer: 100 mM Tris-HCl, pH 7.4 (it should be stored at 4°C).
3. L-α-Phosphatidylcholine (PC) from egg yolk (it should be stored at −30°C).
4. Cholesterol (it should be recrystallized three times from methanol and stored at 4°C under nitrogen).
5. n-Hexane of HPLC grade.

2.3. Measuring System

2.3.1. Film Preparation

1. A Teflon sheet (thickness 12.5 μm).
2. Chloroform of HPLC grade.
3. A marking pin.
4. A DC power supply (10–20 V).

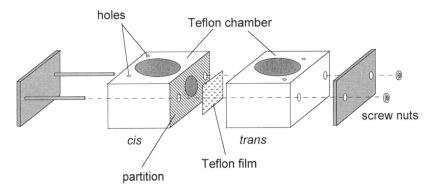

Fig. 2. Schematic illustration of a Teflon chamber used for formation of BLMs by the monolayer-folding method.

2.3.2. Electrode Preparation

1. Two Ag wires (diameter 0.8 mm, length 3 cm).
2. 0.1 *M* KCl.
3. Manicure liquid containing no pearly luster powders.
4. Micropolish alumina powder (diameter 0.3 μm).
5. A dry battery (1.5 V).

2.3.3. Chamber Set Up

1. A Teflon chamber with a partition containing a 6-mm hole that separate *trans* and *cis* compartments (each 1.5 mL) and with four holes for setting electrodes and Teflon tubes (*see* **Fig. 2**).
2. A Teflon film (2- × 2-cm) with an aperture of 100–200 μm in diameter.
3. Two Ag/AgCl electrodes (length 3 cm, diameter 0.8 mm).
4. Two Teflon tubes (length 40 cm, id 0.5 mm) connected to 1-mL disposable microsyringes.
5. Cotton swabs.
6. Silicon vacuum grease.
7. *n*-Hexadecane.

Item 2, Subheading 2.3.1. is also required for this protocol.

2.3.4. Apparatus

1. A patch-clamp amplifier with a preamplifier.
2. A low-pass filter.
3. A personal computer in which pCLAMP software is installed.
4. A chart recorder.
5. An oscilloscope.

2.4. Formation of Receptor-Incorporated Bilayer Lipid Membranes (BLMs)

1. Proteoliposome I: GluR-containing liposomes.
2. Proteoliposome II: Na$^+$/D-glucose cotransporter-containing liposomes.
3. a. Lipid solution I: 8 mg L-a-phosphatidylcholine, dioleoyl (DOPC) and 2 mg cholesterol in 1 mL of chloroform/n-hexane mixture (1:4 volume ratio) (lipid solution should be stored at –80°C under nitrogen).
   b. Lipid solution II: 8 mg PC and 2 mg cholesterol in 3 mL of n-hexane (lipid solution should be stored at –80°C under nitrogen).
4. Bath solution I: 10 μM HEPES/NaOH, pH 7.6, containing 100 mM NaCl, 0.2 mM CaCl$_2$, and 5 mM glycine (it should be filtered through a cellulose nitrate type membrane filter of pore size 0.2 μm).
5. Bath solution II: 100 mM Tris-HCl, pH 7.4 (It should be filtered through a cellulose nitrate type membrane filter of pore size 0.2 μm).
6. 3 M NaCl buffer: 10 mM HEPES/NaOH, pH 7.6, containing 3 M NaCl, 0.2 mM CaCl$_2$, and 5 μM glycine.
7. Concanavalin A: 2.5 mg in 1 mL of 3 M NaCl buffer.
8. Formamide.

   Item 8, Subheading 2.1.1. is also required for this protocol.

2.5. Current Measurement

1. L-Glutamate solutions: 3.0×10^{-9} M; 3.0×10^{-8} M; 3.0×10^{-7} M; 3.0×10^{-6} M; 3.0×10^{-5} M; 3.0×10^{-4} M; 3.0×10^{-3} M; 3.0×10^{-2} M.
2. D-Glucose solutions: 1.5×10^{-7} M; 1.5×10^{-6} M; 1.5×10^{-5} M; 1.0×10^{-4} M; 1.5×10^{-3} M.
3. 1.5 M NaCl.

3. Methods

3.1. Isolation of Receptors

3.1.1. Synaptic Plasma Membranes Containing Glutamate Receptor Ion Channels (GluRs) (see **Note 1**)

All the following steps except for **steps 1–4, 11, 15, 17, 22, 25,** and **27** should be done in an ice bath.

1. Cut the head of an anesthetized rat at the base of the ears with a guillotine.
2. Cut the skull with a nipper and peel off the skull.
3. Take out whole brain by a spatula.
4. Wash with cold water (4°C).
5. Transfer the brain in 15 mL isolation buffer I in a 50-mL beaker.
6. Cut the brain into small pieces (~3 mm) with surgical scissors.
7. Collect the brain pieces from another rat by repeating **steps 1–6**.
8. Homogenize in a 50-mL homogenizer.

9. Transfer the homogenate into a 100-mL beaker.
10. Collect the homogenates from other four rats by repeating **steps 1–9**.
11. Centrifuge at 1480g for 10 min.
12. Collect the supernatant into a 300-mL beaker and store in an ice bath until **step 16** (white pulpy emulsion should not be collected).
13. Suspend the precipitate in isolation buffer I (~6 mL/each centrifuge tube).
14. Homogenize in a 50-mL homogenizer.
15. Centrifuge at 1480g for 10 min.
16. Add the supernatant to the supernatant at **step 12** (white pulpy emulsion should not be collected).
17. Centrifuge at 20,400g for 10 min.
18. Suspend the precipitate in isolation buffer II (~4 mL/each tube).
19. Dilute to 50 mL with isolation buffer II.
20. Hand-homogenize in a 50-mL homogenizer.
21. Layer onto a step gradient of ficoll B, overlayered with ficoll A.
22. Centrifuge at 57,500g for 45 min.
23. Collect the interface between ficoll A and B and transfer in a measuring cylinder.
24. Dilute to 160 mL with isolation buffer II.
25. Centrifuge at 43,700g for 55 min.
26. Suspend the precipitate with isolation buffer II.
27. Centrifuge at 43,700g for 15 min.
28. Suspend the precipitate (synaptic plasma membrane pellets) in a small volume of HEPES buffer.
29. Dilute to 3 mL with HEPES buffer in a spit tube.

3.1.2. Na$^+$/D-Glucose Cotransporter

3.1.2.1. BRUSH BORDER MEMBRANES

All of the following steps except for **steps 1, 3, 4, 9, 14, 15**, and **18** should be done in an ice bath.

1. Open the abdomen of an anesthetized guinea pig.
2. Cut out small intestine and store in 300 MHT buffer.
3. Cut in small pieces (approx 3–5-cm length).
4. Wash the inside of the intestine with cold water.
5. Open the intestine by scissors.
6. Scrape the mucosa by an edge of a thin glass slide.
7. Collect the scrapings with 8 mL 300 MHT buffer in a 100-mL beaker (it should be weighed in advance to know the total weight of the buffer plus the 100-mL beaker).
8. Collect the mucous scrapings from another 9 guinea pigs by repeating **steps 1–7**.
9. Weigh the mucous scrapings.
10. Add an appropriate volume of 300 MHT buffer (8 mL/1 g of the mucous scraping).
11. Homogenize and transfer in a 300-mL beaker.
12. Add CaCl$_2$ to produce its 10 mM concentration.
13. Stir for 20 min.

14. Centrifuge at 2500g for 5 min.
15. Centrifuge the supernatant at 50,000g for 30 min.
16. Suspend the precipitate in a small volume of 400 MHT buffer.
17. Homogenize in a 100-mL homogenizer.
18. Centrifuge at 50,000g for 30 min.
19. Suspend the precipitate in a small volume of 300 MHT buffer and collect in a 500-mL beaker.
20. Add an appropriate volume of 300 MHT buffer to produce 10–15 mg protein/mL (the protein amount should be determined in advance during **step 18**, *see* **Note 2**).
21. Homogenize and transfer in a plastic tube (Ca^{2+}-BBMV) (it should be stored under nitrogen at –80°C until use for **step 1, Subheading 3.1.2.2.**).

3.1.2.2. SDS-Treatment

All of the following steps except for steps **1, 4, 5, 8, 11, 14**, and **16** should be done in an ice bath.

1. Thaw the frozen Ca^{2+}-BBMV at 4°C.
2. Suspend in a small volume of NaCl buffer.
3. Transfer in a 100-mL beaker and add an appropriate volume of NaCl buffer to produce a final concentration of 2 mg protein/mL.
4. Add SDS solution slowly and dropwise under stirring at 4°C to produce its final concentration of 0.02% (v/v).
5. Incubate at 22°C for 15 min under stirring.
6. Cool rapidly in an ice bath.
7. Layer onto a step gradient of sucrose solution II, overlayered with sucrose solution I.
8. Centrifuge at 100,000g for 90 min.
9. Collect the interface between sucrose solutions I and II in a 500-mL beaker.
10. Dilute 10-fold with 25 HT buffer.
11. Centrifuge at 100,000g for 60 min.
12. Suspend the precipitate in 30 mL of KCl buffer.
13. Hand-homogenize the suspension.
14. Centrifuge at 10,000g for 15 min.
15. Transfer the supernatant in a 300-mL beaker and dilute fivefold with 25 HT buffer.
16. Centrifuge at 100,000g for 60 min.
17. Suspend the precipitate in 8–9 mL 300 MHT buffer.
18. Hand-homogenize and transfer in a plastic tube (SDS-BBMV) (it should be stored at –80°C until use for **step 1, Subheading 3.1.2.3.**).

3.1.2.3. Chromatofocusing Chromatography, Dialysis, and Concentration

1. Thaw SDS-BBMV at 4°C.
2. Dilute twofold with pretreatment buffer.
3. Centrifuge at 100,000g for 60 min.
4. Suspend the precipitate in starting buffer to give a concentration of 1 mg protein/mL.

5. Cool in an ice bath for 30 min.
6. Centrifuge at 100,000*g* for 60 min.
7. Apply the supernatant to the top of a chromatofocusing column (the column should be pre-equilibrated with starting buffer).
8. Run the column over pH range of 7.4–4.0 at a rate of 20 mL/min with an elution buffer.
9. Collect the eluted fractions of pH 5.2–5.8 (volume of each fraction ca. 5 mL).
10. Dialyze against 5 HT buffer for 30 h with at least five buffer changes.
11. Concentrate the suspension after dialysis in PVP powder to 2–3 mL with several PVP changes.
12. Subdivide the suspension into small volumes, e.g., 150 μL for each microcentrifuge tube, and store under nitrogen at –80°C (*see* **Note 3**).

Steps 1 and **7–11** should be done at 4°C and **steps 2, 4, 5,** and **12** in an ice bath.

3.2. Preparation of Proteoliposomes

3.2.1. Glutamate Receptor Ion Channel

1. Sonicate the suspension at **step 29** (**Subheading 3.1.1.**) with a tip-type sonicator under nitrogen in an ice bath.
2. Subdivide into small volumes, e.g., 150 μL for each microcentrifuge tube, in an ice bath and store at 4°C under nitrogen (*see* **Note 4**).

3.2.2. Na⁺/D-Glucose Cotransporter

1. Weigh 8 mg PC and 2 mg cholesterol in a pear-shaped flask (*see* **Note 5**).
2. Dissolve with 3 mL of *n*-hexane.
3. Evaporate solvent using a rotary evaporator.
4. Dry under vacuum for 10 min.
5. Add 1 mL of Tris buffer.
6. Agitate using a vortex mixer.
7. Sonicate for 5 min with a bath sonicator.
8. Add 4 μg Na⁺/D-glucose cotransporter.
9. Sonicate for 5 min in an ice bath.

3.3. Measuring System

3.3.1. Film Preparation

1. Cut a Teflon sheet into 2- × 2-cm pieces (*see* **Note 6**).
2. Wash the piece with chloroform.
3. Poke the center of the Teflon film with a marking pin to create a small defect.
4. Wash the film with chloroform.
5. Pass an electric spark through the small defect of the Teflon film (*see* **Note 7**).

3.3.2. Electrode Preparation

1. Polish one end of the Ag wire with sandpaper and then with micropolish alumina powder.

2. Wash with Milli-Q water.
3. Paint the Ag wire except both ends (~5 mm in length) with manicure liquid.
4. Dry in air.
5. Electrolyze in 0.1 *M* KCl for 2–3 h (the Ag wire should be connected to the plus side of a 1.5 V battery).
6. Wash with Milli-Q water and dry in air (*see* **Note 8**).

3.3.3. Chamber Setup (see *Fig. 2*)

1. Wash the Teflon chamber (*trans* and *cis* compartments) with soap.
2. Wash out soap thoroughly with hot water.
3. Wash the chamber with Mill-Q water and dry in air.
4. Paint the outer surface of a partition thinly with silicon vacuum grease.
5. Remove excess grease at the edge of the hole by a cotton swab.
6. Wash the Teflon film with chloroform.
7. Attach on the outer surface of the partition (the aperture of the Teflon film should be positioned slightly above the center of the hole).
8. Tighten the *trans* and *cis* compartments by screwing down the nuts.
9. Paint both sides of the Teflon film thinly with *n*-hexadecane using a cotton swab (*see* **Note 9**).
10. Place the Teflon chamber on a magnetic stirrer in a Faraday cage that is set on a vibration-free table.
11. Set two Ag/AgCl electrodes and two Teflon tubes in the holes of the *trans* and *cis* compartments (the *trans* electrode should be connected to ground and the *cis* electrode to a preamplifier) (*see* **Note 10**).

3.4. Formation of Receptor-Incorporated Bilayer Lipid Membranes (BLMs)

3.4.1. GluR System

1. Transfer 1.24 mL of bath solution I in the *trans* and *cis* compartments of the Teflon chamber.
2. Add 50 µL 3 *M* NaCl buffer to the *trans* and *cis* solutions.
3. Set water level under the aperture of the Teflon film by sucking with micro-syringes connected to the Teflon tubes.
4. Spread 5 µL of lipid solution I on the *cis* and *trans* side solutions (*see* **Notes 11** and **12**).
5. Stand for 10 min for evaporation of solvent.
6. Raise water level slowly and synchronously by operation of two syringes to form a bilayer lipid membrane (BLM).
7. Confirm formation of a BLM by measuring electrical resistance (*see* **Note 13**).
8. Add 50 µL of proteoliposome I to the *cis* side solution.
9. Add 50 µL of HEPES buffer to the *trans* side solution.
10. Add 24 µL formamide to the *cis* side solution.
11. Allow to stand for 30–40 min to fuse proteoliposomes to the BLM.
12. Add 24 µL formamide to the trans side solution.

13. Add 160 μL 3 *M* NaCl buffer to both side solutions.
14. Add 15 μL concanavalin A solution to the both side solutions.
15. Allow to stand for a few minutes for current measurements (*see* **Subheading 3.5.**).

Steps 8–14 should be done while stirring.

3.4.2. Na$^+$/D-Glucose Cotransporter System

1. Transfer 1.50 mL and 1.28 mL of bath solution II in the *trans* and *cis* compartments of the Teflon chamber, respectively.
2. Set water level under the aperture of the Teflon film by sucking with microsyringes connected to the Teflon tubes.
3. Spread 5 μL of lipid solution II on the *trans* side solution and 3 μL of lipid solution II on the *cis* side solution (*see* **Note 12**).
4. Allow to stand for 5 min to evaporate *n*-hexane.
5. Add 220 μL of proteoliposome II to the *cis* side solution.
6. Allow to stand for 5 min.
7. Raise water level slowly and synchronously by operation of two syringes to form a BLM incorporated with Na$^+$/D-glucose cotransporter.
8. Confirm the formation of the BLM by measuring electrical resistance (*see* **Note 13**).
9. Add 10 μL of 1.5 *M* NaCl to the *trans* side solution.
10. Stir the *trans* side solution for 30 s.
11. Allow to stand for 10 min for current measurements (*see* **Subheading 3.5.**).

3.5. Current Measurement (see *Fig. 3 and Note 14*)

3.5.1. GluR System

1. Record background current for 30 s by applying potentials of +30 or +50 mV (*cis* side) (*see* **Note 15**).
2. Add 5 mL of 3.0×10^{-9} *M* L-glutamate solution to the *trans* side solution under gentle stirring (the final concentration of L-glutamate in the *trans* side solution is 1.0×10^{-10} *M*).
3. Stir for 30 s and allow to stand for 5 min without stirring.
4. Record current as in **step 1**.
5. Repeat **steps 3** and **4** after each addition (5 μL) of 3.0×10^{-8} *M*; 3.0×10^{-7} *M*; 3.0×10^{-6} *M*; 3.0×10^{-5} *M*; 3.0×10^{-4} *M*; 3.0×10^{-3} *M*; and 3.0×10^{-2} *M* L-glutamate solutions.
6. Integrate the observed current on a computer (*see* **Note 16**).
7. Plot the integrated currents against L-glutamate concentrations to prepare a calibration graph (*see* **Note 17**).

3.5.2. Na$^+$/D-Glucose Cotransporter System

1. Measure current by applying –50 mV (*cis* side) (*see* **Note 18**).
2. Add 10 μL of 1.5×10^{-7} *M* D-glucose solution to the *trans* side solution (the final concentration of D-glucose is 1.0×10^{-9} *M*).

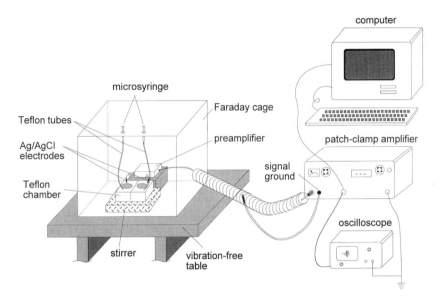

Fig. 3. Experimental setup for current measurements.

3. Stir the *trans* side solution for 30 s.
4. Measure current by applying −50 mV (*cis* side).
5. Repeat **steps 3** and **4** after each addition (10 μL) of 1.5×10^{-6} M, 1.5×10^{-5} M, 1.0×10^{-4} M, and 1.5×10^{-3} M D-glucose solutions to the *trans* side solution.
6. Plot the observed currents against D-glucose concentrations to prepare a calibration graph (*see* **Note 19**).

4. Notes
4.1. Isolation of Receptors

1. Synaptic plasma membranes contain not only GluRs but also a number of other membrane proteins, including voltage-dependent channels.
2. The total amount of protein is determined by Bradford's method using Coomassie Brilliant Blue G-250 dye (available as a Bio-Rad protein assay kit, Hercules, CA).
3. Purified Na^+/D-glucose cotransporter can be used for 3 wk when stored at −80°C.

4.2. Preparation of Proteoliposomes

4. Proteoliposomes containing GluRs can be used for 3 wk, when stored at 4°C.
5. For preparing lipid solution, chloroform or *n*-hexane solution of PC may also be used. The lipid solution is prepared by adding a known volume of PC solution to a preweighed amount of cholesterol in a glass vial. The solvent is then evaporated to dryness under nitrogen stream. Finally, a known volume of *n*-hexane is added.

4.3. Measuring System

4.3.1. Film Preparation

6. A polyimide film can also be used, especially for forming BLMs across an aperture ranging from 20 to 40 μm in diameter. The film may be adopted for recording single-channel currents.

7. An electric spark is generated between a stainless needle and an aluminum plate connected to a dc power supply through a toggle-switch. The small defect on a Teflon film, marked by a marking pin, is positioned just under the needle. The distance between the needle and aluminum plate is to be optimized. The shape and size of the aperture can be observed under a microscope. The film with an aperture of a smooth edge and a circular shape should be used for formation of BLMs. The Teflon film thus prepared can be used for 1–2 mo.

4.3.2. Electrode Preparation

8. It is recommended that two Ag/AgCl electrodes be short-circuited in 0.1 M KCl to age for 1–2 d.

4.3.3. Chamber Setup

9. The Teflon film should be painted thinly with *n*-hexadecane just before use. According to our experience, BLMs are unable to be formed if painting is not performed. On the other hand, excess amount of *n*-hexadecane blocks the aperture of the Teflon film.

10. Physical vibration of the Ag/AgCl electrodes often generates a large frequency noise. To reduce this noise, the electrodes should be tightly positioned in the holes of the chamber.

4.4. Formation of Receptor-Incorporated Bilayer Lipid Membranes (BLMs)

4.4.1. GluR System and Na⁺/ᴅ-Glucose Cotransporter System

11. DOPC is barely soluble in *n*-hexane. The lipid solution is prepared by dissolving DOPC and cholesterol in chloroform and then diluting with *n*-hexane. Chloroform should be used after passing through an Al_2O_3 column.

12. DOPC and PC solutions should be used within 1 wk after preparation. The use of lipid solution after storage of longer than 1 wk results in lowering of the success probability of formation of BLMs or unsuccessful formation of BLMs. Commercially available PC and DOPC solution occasionally contains small particles that are noticeable under light. The use of such lipid solution significantly lowers the success probability of BLM formation.

13. The successful formation of a BLM is known by an increase of resistance from 10 kΩ to 10–250 GΩ. If the formation of a BLM is unsuccessful, **steps 1–7** (GluR system, **Subheadings 3.4.1.** and **3.5.1.**) or **steps 1–8** (Na⁺/ᴅ-glucose cotransporter, **Subheadings 3.4.2.** and **3.5.2.**) are repeated after washing the Teflon chamber only with Milli-Q water, followed by painting the Teflon film with *n*-hexadecane.

4.5. Current Measurement

14. The Teflon chamber and preamplifier should be placed in a Faraday cage on a vibration-free table (an air spring system is recommended). All the shields of the apparatus, i.e., amplifier, filter, and oscilloscope, should be connected and grounded at one place to avoid ground-loop noise. All the connections between the amplifier and filter or oscilloscope should be achieved by using shielded BNC cables.

4.5.1. GluR System

15. The channel currents are recorded with a patch-clamp amplifier. The data, acquired without filtering, are digitized and stored on-line using a computer in which a pCLAMP software was installed.
16. Integrated channel currents are obtained as follows: first, a histogram of the current amplitude is made using a 1-kHz Gaussian cutoff filter on pCLAMP software. Then the event number in the histogram is integrated with respect to a weight of amplitude over the whole amplitude range.
17. The potency of agonists to activate GluRs can be determined by injecting two agonists in series to a BLM. This approach provides a method for evaluating chemical selectivity of GluRs based on a physiologically relevant measure, i.e., membrane permeability changes (*see* **ref. 3**).

4.5.2. Na⁺/D-Glucose Cotransporter System

18. The currents are recorded with a patch-clamp amplifier. Signals from the amplifier are filtered at 3 kHz through a low-pass filter and recorded on a chart recorder.
19. The sensor does not give any response to L-glucose ($10^{-5}\ M$), D-mannose ($10^{-6}\ M$), and D-fructose ($10^{-6}\ M$), but gives a small response to D-galactose ($10^{-6}\ M$).

References

1. Minami, H., Sugawara, M., Odashima, K., Umezawa, Y., Uto, M., Michaelis, E. K., and Kuwana, T. (1991) Ion channel sensors for glutamic acid. *Anal. Chem.* **63,** 2787–2795.
2. Sugao, N., Sugawara, M., Minami, H., Uto, M., and Umezawa, Y. (1993) Na⁺/D-glucose cotransporter based bilayer lipid membrane sensor for D-glucose. *Anal. Chem.* **65,** 363–369.
3. Sugawara, M., Hirano, A., Rehák, M., Nakanishi, J., Kawai, K., Sato, H., and Umezawa, Y. (1997) Electrochemical evaluation of chemical selectivity of glutamate receptor ion channel proteins with a multi-channel sensor. *Biosensors Bioelectron.* **12,** 425–439.
4. Uto, M., Michaelis, E. K., Hu, I. F., Umezawa, Y., and Kuwana, T. (1990) Biosensor development with a glutamate receptor ion-channel reconstituted in a lipid bilayer. *Anal. Sci.* **6,** 221–225.

15

Eukaryotic Cell Biosensor

The Cytosensor Microphysiometer

**Amira T. Eldefrawi, Cheng J. Cao, Vania I. Cortes,
Robert J. Mioduszewski, Darrel E. Menking, and James J. Valdes**

1. Introduction
1.1. The Cytosensor

The Cytosensor, a silicon-based microphysiometer system, is a light-addressable potentiometric biosensor (LAPS) that detects functional responses of living cells in minutes. Thus, unlike other biosensors that detect and quantitate an analyte, the Cytosensor reports on the effect of an analyte on biologic function. It is based on detection of the rate at which cells excrete acidic metabolites, which is closely linked to the rate at which they convert nutrients to energy *(1)*. This is possible because during the experiment the cells are maintained in low buffered medium that contains glucose. This is metabolized in the cells to lactic acid by glycolysis (the most acidifying process in terms of protons produced by turnover of ATP) or oxidized to CO_2 by respiration, with the carbonic acid dissociated and excreted into the medium. Thus, any event that perturbs cellular ATP levels changes energy metabolism and consequently acid secretion. Since cultured cells tend to depend more on glycolysis than on respiration, lactic acid predominates. Thus, there is a direct connection between cellular metabolic rate and the rate of acidification of the medium bathing a cell. The Cytosensor provides a nondestructive method of monitoring cellular acidification rates on a time scale of seconds to minutes *(2)*. A change of 0.01 pH can be detected without any significant change in cell physiology. The sensor is a chip of doped silicon with a thin nitride insulator on the surface that contains hydroxy and amino functions, both of which can be titrated as a func-

From: *Methods in Biotechnology, Vol. 7: Affinity Biosensors: Techniques and Protocols*
Edited by: K. R. Rogers and A. Mulchandani © Humana Press Inc., Totowa, NJ

tion of pH. A photocurrent is generated when the sensor is illuminated with an amplitude-modulated light-emitting diode. A change in the surface potential can be measured by determination of the changes in an applied potential required to produce a photocurrent. Therefore, the silicon chip detects the surface potential at the light-activated region of the interface between a salt solution (low-buffered medium) and the insulated surface of the chip. Since the surface of the sensor reversibly binds protons, its surface potential is pH-dependent in a Nernstian fashion.

In the flow chambers, cells are retained between two microporous (proton-permeable) polycarbonate membranes in a flat cylindrical region (50-mm high, 6-mm diameter). This small volume of 1.4 μL serves several functions. First, it contributes to high sensitivity measurements by limiting the volume, and therefore buffer capacity, into which acidic metabolites are excreted. Second, it limits the time required for diffusion of metabolites and nutrients across the chamber. This produces a homogeneous distribution of molecules of interest within the chamber on the timescale of the measurement, insuring that all of the cells in the measurement region are in an equivalent chemical environment. A third advantage is that only small quantities of precious reagents and test agents are required to test the cellular response. Adherent cells can be grown directly on the cell cups; nonadherent cells can be immobilized by methods including a thin fibrin clot. The cell capsule is sealed between a LAPS chip (below) and a plunger (above) that supplies culture medium via a superfusion system (**Fig. 1**). A light-emitting diode (LED) positioned below the LAPS enables pH measurements on the central 2-mm diameter circular region of the cell capsule.

The measurement of cellular acidification rate begins with loading the capsule cups seeded with the cells into the Cytosensor chamber, starting the flow of medium, then interrupting it for 30 s, which reduces the chamber pH. The cellular acidification rate is determined from a linear fit of the pH versus time data. The flow is then resumed, resetting the pH of the chamber to that of the fresh medium. This process is repeated continuously for a period of minutes to hours depending on the experiment. Test samples are prepared in disposable plastic centrifuge tubes from which they are automatically pumped to the different cell flow chambers during the course of an experiment. Between sample introductions, the cells are perfused with drug-free medium to remove the test chemical. There must be a dose–response relationship. A drug like acetaminophen (a known hepatotoxicant at high doses) may increase acidification rate (e.g., by human hepatocytes) at low doses (e.g., 5 m*M*) and inhibit it at higher ones (e.g., 30 m*M*) *(3)*.

The Cytosensor has eight identical channels, each consisting of a disposable sterile capsule (in a thermally controlled environment) that accommodates

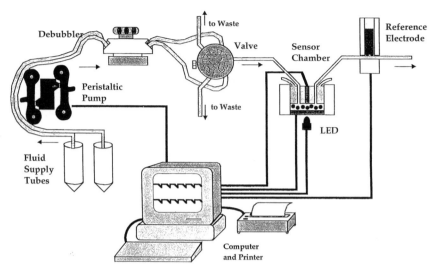

Fig. 1. Schematic diagram of the Cytosensor showing the media flow in one channel, with fluid supply tubes, pump, debubbler, valve, the sensor chamber, and Ag/AgCl reference electrode that is coupled to the chamber exit tube and computer. Not shown is the circulating water bath that controls the temperature of the flow chamber (with permission from Molecular Devices).

eukaryotes and prokaryotes. The multiple channels of units allow for running a whole experiment simultaneously. One unit can be used for control cells and each of the other seven for a different concentration of the test chemical. Alternatively, after the baseline metabolic rate is determined, increasing concentrations of the same chemical can be added sequentially to the same cells in one channel. The concentration of test material required to reduce or increase the acidification rate to 50% relative to the baseline is computed by interpolation between the rate data points spanning the 50% response level. Instrument control, such as medium flow rate (pump speed) and valve-switching, as well as data acquisition, analysis, and storage is handled by an attached computer and software. The instrument can be programmed to make measurements every 30 s to hours and for up to 5 d. Thus, not only can cell metabolism be monitored during exposure to various concentrations of a drug, but also after wash with buffer to determine the reversibility of the changes observed and cell death.

Different cell cultures are used based on the research need (e.g., human liver or neuronal cell lines), and these may require different conditions (e.g., media, dissociation solutions, handling, cell density). Also, different protocols for exposure of the cells in the Cytosensor are used based on the objectives of the research. If the need is for observing functional neurotransmitter receptor–

ligand interaction, such as the case in activating excitatory muscarinic or adrenergic receptors that produce membrane depolarization, the increases in acidification rates are transient and reversible and may be observable for a few seconds *(1,2)*. This requires frequent measurements of acidification rate, which also applies to detection of the activation of inhibitory γ-aminobutyric acid receptor that produces membrane hyperpolarization and reduced metabolism. On the other hand, if the need is to study the effect of a nerve growth factor on cell proliferation, measurements every 2–5 min suffice. A complete dose–response function for receptor activation may be completed in 30–60 min. Alternatively, if the interest is in studying the toxic effect of a drug, in which the change may be long term and irreversible, a short protocol for 1–2 h or a long one for 24 h with measurements every 1–5 min are used. The long exposure is more accurate for predicting basal cytotoxicity of a drug to most human cells, i.e., nonspecific cytotoxicity to cellular functions and structures that are basic to most cell types. We have found excellent correlations between the 24 h in vitro cytotoxicity data on 10 drugs on human liver cells generated by the Cytosensor and a fluorescence assay using the fluorescent Calcein AM dye ($r = 0.991$) and the [^{3}H]thymidine uptake assay ($r = 0.976$) on the same cell line and their published human lethal blood concentrations ($r = 0.958$) *(3)*. The paucity of information available on human toxicity to provide a database for regulatory purposes is a driving force for developing new methods and applying new technologies to in vitro toxicology research. This is needed to supplement or replace in vivo animal testing.

2. Materials
2.1. Cell Culture
2.1.1. Cell Lines

1. Human liver cells (CCL13, American Type Culture Collection (ATCC), Rockville, MD).
2. SK-N-SH (human neuroblastoma, ATCC HTB-11).
3. PC 12 cell line (from NIH, Bethesda, MD).
4. SH-SY5Y cell line (cloned neuroblastoma cells from Stephen Fisher, University of Michigan, Ann Arbor).
5. Rat liver cell, clone 9 (ATCC CRL-1439).

2.1.2. Apparatus

1. Laminar flow biosafety hood equipped with HEPA filters, vacuum device, and burner.
2. 37°C water bath.
3. 37°C incubator with CO_2 tank adapted.
4. Inverted microscope (at least 10× magnification).
5. Refrigerator/freezer and liquid nitrogen tank.

6. Hemocytometer/cell counter.
7. Centrifuge—up to 600*g* speed.

2.1.3. Labware

1. Sterile plastic and glass pipets and tips.
2. Pipet helper (automatic aspiration/dispensing device).
3. Sterile culture flasks, centrifuge tubes, and cryostorage tubes.

2.1.4. Media, Dissociation Solutions and Growth Factors

1. See **Table 1**.
2. Agarose entrapment medium for nonadherent cells (MDC, Sunnyvale, CA).

2.1.5. Disinfectant Solutions

1. Alcohol (70%), bleach, germicidal soaps.

2.1.6. Lab Coats, Surgical Gloves

1. Standard for laboratory use.

2.2. Cytosensor

1. Cytosensor parts (**Fig. 1**): Cell capsules comprise the following: capsule cups, spacers and inserts, the sensor chambers and sterilized Cytosensor System (Molecular Devices, Sunnyvale, CA).
2. Running buffer.
3. Media: Different depending on the cells and experiment (**Table 2**): DMEM (Gibco, Grand Island, NY); F12 (Gibco), RPMI (MDC), MEM Earle's salts, nonessential amino acid (Gibco), 4 *M* NaCl for all except RPMI. Approximately 1000 mL of medium, pH 7.3–7.4, and sterilized by filtration (0.2 μm) (*see* **Note 2**).
4. Labware:
   a. Sterile 50-mL and 15-mL centrifuge tubes (VWR).
   b. Sterile plastic centrifuge tubes (50 mL) for cleanup.
   c. Forceps.
   d. 16 (50-mL) centrifuge tubes for running media.
5. Drugs and toxicants of interest.
6. Ethanol: Ethanol (70%) is used for cleaning a variety of Cytosensor surfaces and fluid paths.
7. Cytosensor sterilant (Molecular Devices Co.): Cytosensor sterilant is used after each Cytosensor experiment to sterilize and depyrogenate without corrosion.

3. Methods

3.1. Cell Culture

3.1.1. Handling Frozen Cells

1. Initiate culture as soon as possible upon receipt. Record the batch, date and cell passage number. Keep extra tubes in liquid nitrogen (*see* **Note 1**).

Table 1
Cell Culture Conditions

Cell type	Media	Dissociation solution	Cell density (Cytosensor)
Human liver cell line or rat liver cell line	Growth media: Basal medium eagle (BME—Gibco) with 10% calf serum—pH 7.4. Freeze media: Growth media + 5% DMSO	4 mL of trypsin/EDTA solution (0.25% trypsin + 0.03% EDTA in BME with 10% calf serum). Leave in contact for 5 min (until detached).	2×10^5 cells/cup
PC 12 Cell line [a]	Growth media: Dulbecco's modified eagle medium (DMEM—Gibco) with 7.5% fetal bovine serum + 7.5% horse serum. Freeze media: Growth media + 5% DMSO	8 mL of Dulbecco's phosphate buffer without Ca^{2+} and Mg^{2+}. Leave in contact for 10 min (bumping might be necessary to promote detachment).	2×10^5 cells/cup
Human SK-N-SH cell line [b]	Growth media: Minimum essential media (Eagle—Gibco) with 1 mM sodium pyruvate and 0.1 mM nonessential amino acids and Earle's BSS + 10% fetal bovine serum Freeze media: Growth media + 5% DMSO	4 mL of trypsin solution (0.25%) in growth media. Leave in contact for 5 min until detached.	3×10^5 cells/cup
Human SH-SY5Y cell line [b]	Growth media: Dulbecco's modified eagle media—high glucose + 10% fetal bovine serum (Sigma) Freeze media: growth media + 10% fetal bovine serum + 10% DMSO	8 mL of Dulbecco's phosphate buffer without Ca^{2+} and Mg^{2+}. Leave in contact for 10 min until detached.	2×10^5 cells/cup

[a]Culture flasks should be collagen-coated.
[b]These cells require media renewal every 3 days.

Table 2
Running Media Composition

Running media	Possible supplier	Media in pkg (1 L)	NaCl (4 M)
DMEM	Gibco Cat.# 12800-017	1	11.1 mL
F12	Gibco Cat.# 21700-075	1	3.6 mL
RPMI	MDC Cat.# R8016	1	—
MEM Earle's salts, nonessential amino acids	Gibco Cat.# 41500-034	1	6.6 mL
F12/DMEM	Gibco Cat.# 12500-062	1	7.31 mL

2. Thaw the cell tube by rapid agitation (40–60 s) in 37°C water bath.
3. When ice is melted, immerse the tube or wipe it with 70% ethanol at room temperature.
4. Use aseptic conditions: Pass the bottles', tubes', and flasks' necks through a flame whenever opening or closing caps. Work within the flame area. Before bringing any item into the laminar flow hood, wipe each with 70% alcohol. Have all the needed labware available inside the hood for the following operations.
5. Inside the hood, open the tube (within flame area) and withdraw contents into a 15-mL sterile centrifuge tube using a sterile glass pipet.
6. Add 6–8 mL of the required media (**Table 1**) and homogenize gently.
7. Centrifuge the above diluted cell suspension at 300g for 5 min.
8. Inside the hood, carefully discard the supernatant.
9. Resuspend the pellet, adding 3–5 mL of the recommended growth media and homogenize gently up and down with a sterile pipet attached to the pipet helper.
10. Transfer the cell suspension into sterile culture flasks.
11. Add the recommended growth media (**Table 1**) to each of the culture flasks to obtain a total of 15–20 mL in each 75-cm^2 culture flask.
12. Gently agitate each flask and label with the passage number, date, and cell type.
13. Place the flasks in a 37°C incubator at a 5–10% CO_2 atmosphere.
14. After incubating overnight, check for cell adherence to flasks.
15. On subsequent days, check daily by microscope for cell conditions (e.g., absence of contamination, confluence). Replace media whenever floating cells are noticed and proceed to a new passage (*see* **Subheading 3.1.2.**) whenever the cells reach confluence.

3.1.2. Handling Cell Passage (Confluence of Cultured Flask)

1. Warm up the required growth media and dissociation solution as described in **Table 1** in a 37°C water bath.

2. Remove the confluent flask(s) (where flask bottom is fully covered) from the incubator.
3. Have all the needed labware available inside the laminar flow hood.
4. Aspirate the whole media from each flask, using a glass Pasteur pipet attached to the vacuum line in the hood.
5. Add the indicated amount of dissociation solution (**Table 1**).
6. Leave in contact for the required time and verify that the bottom wall of each flask has no cells attached.
7. Collect the cell suspension into a 50-mL centrifuge tube.
8. Rinse the flask with about 8 mL of media and add the rinse to cells in the centrifuge tube.
9. Centrifuge the cell suspension at $300g$ for 5 min.
10. Resuspend the pellet as described in **Subheading 3.1.1.** For cell passage the subculturation ratio is usually 1:8.
12. Count the cells under a microscope using a hemocytometer plate and determine the density/mL.
13. Rehomogenize the cell suspension up and down with the sterile pipet.
14. Make the necessary dilutions with the required media according to the desired density in **Table 1**.
15. Reserve enough quantity of the cell suspension ($\approx$10 mL) to perform the experiment.

3.1.3. Media Renewal Between Passages

This procedure applies only for SH-SY5Y cells and is done every 3 d as follows:

1. Warm up (in 37°C water bath) the recommended media (10 mL × number of flasks that need media renewal).
2. Take the culture flask(s) out of the incubator and place in the hood.
3. Work under aseptic conditions (within flame area, passing flasks' necks through flame).
4. Aspirate 10 mL of each flask media with a sterile 10-mL pipet. Discard.
5. Transfer 10 mL of new media into each flask using a sterile pipet.
6. Place the flask(s) back in the 37°C incubator.

3.2. Test Compound Dilutions

1. Prepare serial 10–100-fold dilutions of stock test solutions with sterile, low-buffered medium after equilibrating to room temperature using sterile 10-mL pipets and an automatic pipetor. The total volume of test material required per dilution (up to eight) and per channel is as follows: Using 50% maximal pump speed (0.1 mL/min), multiply total exposure time (30 min/dilution) by five dilutions. Therefore, 15 mL of test material will be needed per channel for a typical experiment.

3.3. Cytosensor Protocols

3.3.1. Assembling the Cell Capsules

All assembling steps should be done in a tissue culture hood using aseptic techniques. The Cytosensor system uses disposable cell capsules to immobilize living cells in the sensor chamber. The cell capsule comprises the capsule cup, which provides support for the cells; the spacer, which defines the height and diameter of the compartment in which the cells are confined; and the insert, which protects the cells from being damaged by the flow of medium through the chamber and also prevents them from washing out of the chamber.

3.3.1.1. Capsules with Adherent Cells Grown in the Cups

1. Just before use, put a spacer in each capsule cup, lowering it into the cup at an angle to avoid trapping bubbles underneath it. Gently push the spacer under the medium until it rests on the bottom of the capsule cup. Avoid pushing on the cup's membrane, because it tears easily. Check that there are no bubbles trapped under the spacer, gently lifting the spacer to free bubbles if necessary.
2. Put a dry capsule insert into each capsule cup. Never use a wet insert, since it will trap bubbles in the capsule and will not sink. Pipet a few drops of sterile medium into each insert to help it sink, then wait a few minutes for the insert to sink at its own rate without pushing down to speed up the process. When fully sunk, the rim of the insert will protrude slightly above the cylindrical portion of the cup.
3. When the insert has sunk to the cup's bottom check the capsule for bubbles trapped between the cup and the insert. If a capsule contains a bubble, gently lift out the insert, remove the bubble, then put in a new dry insert.
4. The cell capsules are now ready to be loaded into the sensor chambers. Do not let the assembled capsules stand for more than 15 min and do not put them in an incubator, since bubbles can form inside the capsules.

3.3.1.2. Capsules with Nonadherent Cells and Fibrin Matrix

1. Use dry capsule cups without spacers.
2. Prepare cell suspension at the recommended desired density ($2 \times 10^6/150$ μL).
3. Mix the cells and molten agarose (at 37°C) at a 3:1 ratio.
4. Deposit a 7-μL drop of agarose/cell mixture at the center of the dry cup.
5. Let the drop form a gel for 5–10 min (check the state of the drops in the cups by pipeting a control drop on the side of the cup).
6. Pipet 2 mL of running medium outside the cup and 500 μL of running medium in each cup.
7. Place a spacer in each cup: using forceps, touch the surface of the medium with the spacer at an angle and drop the spacer in the cup. Do not touch the agarose/cell drop with the spacer; hold the spacer at an angle before dropping it in the cup.
8. Using forceps, gently tap on the spacer until it sinks to the bottom of the cup.
9. Place an insert into each cup.
10. Sink the insert by pipeting approx 1 mL of running medium into each.
11. Place the assembled capsules into the sensor chambers.

3.3.2. Loading and Installing the Sensor Chambers

1. Working in a tissue culture hood, under aseptic conditions, aspirate the equilibration medium from the sensor chambers, then pipet 2 mL of the low-buffered

medium into each chamber. Transfer a cell capsule into each sensor chamber, lowering the capsule into the chamber at an angle to avoid trapping bubbles under it. Cover each sensor chamber with half of a sterile Petri dish (this step is optional, to help keep the chambers sterile), then carry the chambers to the Cytosensor (*see* **Note 3**).

2. Set a sensor chamber beside one of the cleaning chambers on a workstation. Crank the gantry lever back to raise the rod, then swing the gantry to the left, away from the cleaning chamber. Grasp the plunger by the rubber handle and lift it out of the cleaning chamber, being careful to keep a drop of medium hanging from the bottom of the plunger. If the O-ring has fallen out of the plunger, press the plunger down over the O-ring in the cleaning chamber until the O-ring is firmly seated in the circular groove in the bottom of the plunger.

3. While holding the plunger, remove the cleaning chamber and put the sensor chamber on the base pad, aligning the yellow dot on the chamber with the yellow dot in front of the base pad.

4. Gently lower the plunger into the cell capsule in the sensor chamber, again being careful to keep a drop of medium hanging from the bottom of the plunger. As you lower the plunger, align the ridge on the plunger skirt with the mark on the upper rim of the sensor chamber, so that the tabs on the underside of the plunger skirt fit into the notches on the chamber's inner rim. Avoid twisting or replacing the plunger after it has been seated in the chamber, since that can dislodge the cells or the entrapment matrix. Swing the gantry into position over the plunger, then crank the lever forward to lower the rod.

5. Repeat **steps 2–4** with the remaining sensor chambers, then check for bubbles in each chamber's outlet tubing and reference electrode. If you find a bubble, make sure it is moving with the fluid stream, and eventually flows out past the reference electrode. If a bubble sticks to the tubing or in the electrode, gently flick the tubing or electrode with your finger or a pencil to encourage the bubble to move. If it still remains stuck, detach the outlet tubing from the plunger outlet and blot the end of the tubing back on the plunger, then blot up any liquid that has dripped from the plunger. The large bubble should sweep any smaller bubbles out of the fluid path.

6. After having installed all the sensor chambers and plungers on the instrument and removed all visible bubbles, close the workstation's lids.

3.3.3. Cytosensor Operation

3.3.3.1. Setting the Parameters

1. To set the **Preferences**, go to the **Edit** menu and click on **Preferences**.

2. In the dialog box displayed, specify how much of the data should be saved. Next to the **Data Retention** prompt, click on **All** to save all of the raw data or click on **Last Half Hour** to save only the last 30 min of raw data. (The instrument will run out of memory if you try to save all the data from an overnight experiment).

3. Set the temperature at which the experiment is going to be run (typically 37°C) and the temperature difference between the debubblers and the chambers (typically 6°C higher).

4. After entering all the information, click OK.
5. To set the pump cycle, go to Control menu and click on Edit Pump Cycle.
6. In the dialog box displayed, enter a time for the first interval (how long the pumps stay on).
7. Enter a time next to the Rate Measurement prompt (how long the pumps should be off and the rate measured). Allow 3–5 s after the pump stops before starting the rate measurement. During this short delay, the flow may not be completely stopped yet, causing a slight shoulder in the raw data. Typically, there will be a 2-s interval between the end of the rate measurement and the beginning of the next pump cycle. This delay allows avoidance of any possible interference caused by the pumps being turned back on.
8. Enter the Total Cycle Time (how long is one complete cycle).
9. After entering all the information, click OK.

3.3.3.2. FEEDING THE FLUID PATHS

1. Fill 50-mL sterile centrifuge tubes with running buffer, warmed up at 37°C in water bath, and place on workstations.
2. Using the lower portion of the Front Panel, set the pump speed to 50% for at least 5 min, switching valves periodically.
3. Using the Front Panel, set the pump speed to Idle.

3.3.3.3. DATA COLLECTION

1. Once all visible bubbles have been removed, close workstation lids. Click Run on the Front Panel.
2. A dialog box indicates that the instrument is calibrating the workstations.
3. When calibration is complete, data collection begins. The Now Bar will move to the right as each data point is displayed on the screen.
4. If there is a problem with calibration a dialog box will indicate which channel(s) cannot be calibrated. Click on the Cancel button to stop calibration, click on the Stop button in the Front Panel to stop the experiment, then look for the source of the problem and refer to the Basic Troubleshooting section for how to proceed.

3.3.3.4. BASELINE AND CONTROL SETTINGS

1. When certain cells are put in the Cytosensor, they require a period of time to equilibrate, as evidenced by a gradual change in the rate data. It is important to let the rates settle to a steady baseline before introducing the compound of interest. If you introduce your test solution before the rates are steady, you may be unable to distinguish whether a change in rate is caused by the test solution or is a continuation of the equilibration process.
2. Before putting the test solution in the system, at least one control switch should be performed. This means changing the valve so the media that perfuses the cells comes from side 2 instead of side 1. The control switch verifies that the pH of the buffer on both sides is the same and that switching the valve does not, by itself,

affect the rate data. Little or no change in the rate data should be seen when a control valve switch is performed. To perform a control valve switch, add a fluid interval (in **step 4** a span of time that a valve is in either position 1 or 2—the valve position determines the fluid path routed to the chamber) as described *below*.

3. The valve schedule for each chamber is a chart of the fluid intervals that is displayed by default under the rate data, and can be displayed beneath the raw data. To display the valve schedule under the raw data, go to the Screen menu, hold Valve Schedule, and select All, None, or Displayed Traces. It may be useful to visualize the pump schedule on the rate data screen, since the valve schedule sometimes needs to be coordinated with the pump schedule. To display the pump schedule under the rate data, go to the Screen menu and select Pump Schedule.

4. To add a fluid interval, make sure the rate data is the active window, go to the Controls menu and highlight Add Fluid Interval. Drag the cursor to one of the options on the submenu displayed. If you choose Displayed Schedules, a fluid interval will be added to all of the displayed valve schedules. In the Fluid Interval dialog box, you can enter a name and a concentration in the text box next to the Fluid Name prompt. If you are creating a control valve switch, you might want to name it "control".

5. Enter the desired start time and duration next to the Interval Start and Interval Duration prompts. After that, click OK. The fluid interval will then be added and displayed on the valve schedule. It is not possible to create a fluid interval if the start time is in the past relative to the Now Bar.

3.3.3.5. Challenging the Cells with a Compound of Interest

1. Once the rates have settled to a steady baseline and you have performed a control valve switch, put the tubes containing the compound of interest on side 2 and add a fluid interval (*see above*). Side 1 is perfused with control media.

2. Be sure to allow at least three pump cycles to elapse between when the drug is put in side 2 and when the valve change occurs. This time allows the drug to travel along the fluid path and reach the valve.

3. Typically, if the exposure time is shorter than one pump cycle, the interval start will be chosen so that the end of the exposure coincides with the end of a pump cycle. If the exposure time is as long as, or longer than, one pump cycle, the exposure time typically coincides with the beginning and the end of a pump cycle.

4. When the valve switch occurs the fluid entering the cell chamber will be coming from side 2, exposing the cells to the compound of interest.

3.3.3.6. Exposure Time Protocols for Detection of Cytotoxicity

1. The short (4-h) protocol: Introduce the drug at the lowest desired concentration into the flow medium with the instrument taking measurements every min for 30 min. Double the drug concentration further and after 30 min repeat this step until five higher doses are tested. Flow drug-free medium for 10 min to determine the reversibility of the effect of the drug (*see* **Notes 4** and **5**).

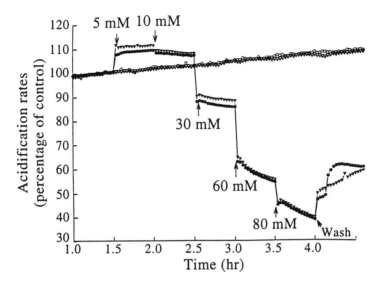

Fig. 2. Data obtained from a 4-h protocol on the effect of acetaminophen (5–80 m*M*) on the metabolic activity of human liver cells. The cells were exposed to sequentially increasing concentrations for 30 min each, then recovery was monitored for 30 min after removal of acetaminophen (wash). The acidification rates of untreated cells in two channels (△,○) were used to calculate the effect of exposure to the drug on cells in two other channels (●,▼). (with permission from **ref. 3**).

2. The long (24-h) protocol: Expose the cells to the drug for 20 h, while the cells are plated in the capsule cups. Use one channel as control (drug-free) and each of the other seven channels with an increasing dose of the drug in the perfusing medium. Program the computer to record acidification values every 6 min for 4 h. Perfuse the cells with drug-free medium to record reversibility of the effect (*see* **Fig. 2**).

3.3.3.7. NORMALIZING THE DATA

1. Although the rate data from all eight replicate chambers are likely to be similar, they are not likely to be identical. To compare the responses of the various cell chambers with one another, it is necessary to "normalize" the data. This function will set all the baselines to 100%, allowing the changes in acidification rates to be comparable.
2. To normalize the data, highlight an area of the rate data that is representative of the baseline before the compound of interest is introduced. Click and hold down the mouse button at one end of the region to be highlighted. Drag the cursor across and release the mouse button when the entire region of interest has been highlighted.
3. In the Screen menu, click on Normalize (or use the keyboard shortcut, <><B>).

4. Cytosoft will calculate the average value of each data trace (i.e., the data from a single chamber) during the selected period, then normalize the data in each trace to the baseline value.
5. The normalized data will be displayed as a percent of the baseline value, with the baseline value as 100% (**Fig. 2**).
6. When the experiment is finished and cleaning is ready to start, save the experiment (select the Save in the File menu) and close the file (select Close File in the File menu).

3.3.4. Cleaning and Sterilizing the Cytosensor

The Cytosensor system requires some routine maintenance to keep it in good working order. The system should always be either sterilized for the next use or prepared for storage. The Cytosensor sterilant was developed specifically to sterilize and depyrogenate the Cytosensor (*see* **Subheading 2.2.6.**).

1. Replace tubes of running buffer with tubes of deionized H_2O. Allow water to flow through the system for at least 10 min, at high flow rate (90%), or until all medium is flushed out of the system. It is important to flush the medium out of the plungers and tubing to prevent salt crystallization.
2. Switch valves at least once. To manually switch valves, go to the Front Panel. If the valves are currently set to side 1 (all the dots are under the number 1), click on the number 2. If the valves are set to side 2, click on the number 1.
3. Prepare sterilant (this solution has to be used within 2 h after the preparation), by adding 75 mL 4X solution A + 225 mL sterile deionized H_2O + 600 μL solution B. The 1X solution A can be prepared in bulk and stored at room temperature. Dispense 14 mL of the sterilant solution into each of 16 (15-mL) conical tubes.
4. Using Control Panel, set Pump Speed to 0%.
5. Unscrew tops of debubblers and discard used membranes.
6. Using lab wipes, clean each debubbler with deionized H_2O (to dissolve crystals) followed by 70% ethanol (to clean and dry). Make sure the bottom portion of the debubbler (the heating pad) is perfectly clean and dry. Put a fresh membrane on each debubbler. Replace tops, screwing them down securely, so they are finger tight.
7. Remove used chambers and plungers and replace with autoclaved cleaning chambers and plungers: lift the gantry rod and move it to the left, place the sensor chambers to the right of the LED site, place the cleaning chambers above the LED site; and move and lower the gantry back in place. Disconnect the inlet and outlet tubings from the old plunger and connect them to the fresh one. **Attention:** Cleaning chambers and plungers are autoclaved, sensor chambers cannot be autoclaved.
8. Remove water tubes, put the 15-mL conical tubes containing sterilant into empty 50-mL holder tubes and put sterilant tubes onto the system.
9. Using Control Panel, set Pump Speed to 50%. Manually switch valves every 2–3 min for at least 10 min, then using Control Panel, set Pump Speed to 0%.

10. Fill each of 16 (50-mL) tubes with 45 mL of sterile deionized H_2O, then place them in the cytosensor after rinsing the intake tubing.
11. Double click on Cytosoft Cleanup Protocol. Under the Edit menu, click on Preferences. In the Preferences dialog box: set Default Pump Speed Idle to 0.5%, set Data Retention to Last Half Hour, set the chamber temperature to 20°C and the chamber to debubbler temperature difference to 0, and click OK.
12. Close file (do not save changes). Double click on Cytosoft Cleanup Protocol to reopen the file.

3.3.5. Sterilizing the Sensor Chambers

1. Remove cell capsules from the sensor chambers and discard capsules in a biohazard waste container. Thoroughly rinse the inside of the chambers with sterile distilled water, scrubbing them gently with a swab to remove dried medium if necessary.
2. Swab the chambers with a lab wipe soaked in 70% ethanol. Be sure to wipe the top rim of the chambers with the ethanol. If sterilizing immediately, transfer the chambers to the biosafety hood. If sterilizing later, dry the chambers and store them until needed.
3. Fill each chamber with 5 mL sterilant. Using a swab, gently spread sterilant around the small inner rim of the chamber. Be careful not to get sterilant on the top rim of the sensor chamber, especially the chamber contact. If sterilant gets on the contact, wipe it with a dry swab. Allow the chambers to soak in the sterilant for a minimum of 2 h or overnight.
4. Thoroughly aspirate the sterilant from the sensor chambers, using a sterile plastic pipet to avoid scratching the chamber. Fill the chambers with 6 mL sterile water and carefully swab them. Aspirate the water from the chambers, refill with water, and aspirate at least 5 more times. Finally, fill with 5 mL sterile water and cover until the next use.

4. Notes
4.1. Cell Culture

1. Once the cell culture is in progress, part of the initial passages should be stored in liquid nitrogen as a stock standard for comparison and emergency purposes (e.g., contamination or mutation occurring from passage to passage). In this case, when obtaining the cell suspension, transfer 1 mL of a 10^6 cells/mL suspension to a cryostorage tube and add protective freezing additives, such as 5–10% dimethylsulfoxide (DMSO) and additional percent of growth factor as described under freeze media (**Table 1**).

4.2. Cytosensor

2. Because the Cytosensor measures the ability of the cells in the chamber to change the pH of their medium, the running medium must have a low-buffering capacity (typically 1 m*M*). To reduce the buffering capacity and avoid forming bubbles in

the sensor chamber, sodium bicarbonate should be omitted from the medium and sodium chloride substituted to preserve osmotic balance. Typically, the running buffer is a serum-free growth medium, without bicarbonate. If necessary, the running buffer may be supplemented with human or bovine serum albumin (1 mg/mL).

3. All assembly steps should be done in a tissue culture hood using aseptic technique. Do not begin capsule assembly unless you will be able to install the capsules on the microphysiometer workstation within 30 min. Be careful not to trap bubbles inside the capsules for they will cause extremely noisy data.

4. Acidification rate can be expressed in terms of H^+/s per cell. However, when comparing various mammalian cells, these vary tremendously in size depending on length of time in culture (up to 2×) and cell type. They range from an epithelial cell of 1–4 pL to a large neuron with 1-m long axon and 20-μm in diameter (i.e., 300 nL). It is best in this case to express the acidification rate relative to cell protein as H^+/s per ng protein, which generally ranges from 0.1 to 1.3 ng/cell.

5. If washing the cells in the Cytosensor with drug-free medium does not totally reverse the drug's effect, the percent reduction reflects either a reduction in metabolic activity of all cells, death of some, or a combination of both. If a reduction is irreversible by washing in drug-free media, then this is most probably caused by cell death.

Acknowledgment

This research was financed partly by Intergovernmental Personnel Act to A. Eldefrawi from US Army and NIH T32 # ESO7263 grant.

References

1. McConnell, H. M., Owicki, J. C., Parce, J. W., Miller, D. L., Baxter, G. T., Wada, H. G., and Pitchford, S. (1992) The Cytosensor microphysiometer—biological application of silicon technology. *Science* **257,** 1906–1912.

2. Owicki, J. C. and Parce, J. W. (1992). Biosensors based on the energy metabolism of living cells: the physical chemistry and cell biology of extracellular acidification. *Biosens. Bioelectron.* **7,** 255–272.

3. Cao, C. J., Mioduszevski, R. J., Menking, D. E., Valdes, J. J., Cortes, V. I., Eldefrawi, M. E., and Eldefrawi, A. T. (1997) Validation of the Cytosensor for *in vitro* cytotoxicity studies. *Toxicol. In Vitro* **11,** 258–293.

Index